The Citizen Marketer

Oxford Studies in Digital Politics

Series Editor: Andrew Chadwick, Royal Holloway, University of London

The Citizen Marketer

PROMOTING POLITICAL OPINION IN THE SOCIAL MEDIA AGE

JOEL PENNEY

OXFORD
UNIVERSITY PRESS

OXFORD
UNIVERSITY PRESS

Oxford University Press is a department of the University of Oxford. It furthers
the University's objective of excellence in research, scholarship, and education
by publishing worldwide. Oxford is a registered trade mark of Oxford University
Press in the UK and certain other countries.

Published in the United States of America by Oxford University Press
198 Madison Avenue, New York, NY 10016, United States of America.

Library of Congress Cataloging-in-Publication Data
Names: Penney, Joel, author.
Title: The citizen marketer : promoting political opinion in the social media
age / Joel Penney.
Description: New York : Oxford University Press, 2017. |
Series: Oxford studies in digital politics |
Includes bibliographical references and index. |
Identifiers: LCCN 2016047946 (print) | LCCN 2017012226 (ebook) |
ISBN 9780190658076 (Updf) | ISBN 9780190658083 (ePub) |
ISBN 9780190658090 (online) | ISBN 9780190658052 (hardback) |
ISBN 9780190658069 (paperback)
Subjects: LCSH: Political participation—Technological innovations. |
Communication in politics—Technological innovations. |
Social media—Political aspects. | Marketing—Political aspects. |
Public opinion—Political aspects. | BISAC: SOCIAL SCIENCE / Media Studies.
Classification: LCC JF799.5 (ebook) | LCC JF799.5 .P46 2017 (print)
DDC 324.7/3—dc23
LC record available at https://lccn.loc.gov/2016047946

To my parents and grandparents, for teaching me the value of books
and putting a pen in my hand

Contents

Acknowledgments

This book is the culmination of more than eight years of writing and research, and it would not have been possible without many wonderful colleagues and friends who have helped me along the way. First, I owe an immeasurable debt to Barbie Zelizer, my graduate advisor, who first encouraged me to explore citizen political expression at a time when my scholarly interests were varied, to say the least. I vividly remember the day when I sat in her office discussing my fascination with graphic T-shirts, and she urged me to focus on the political ones, since those would likely be the most interesting to study. Thus began a long and winding journey that took me from photoethnography projects at presidential campaign rallies to the wide and complex world of political social media, resulting in the constellation of research featured in this book. Not only did she inspire my intellectual pursuits for years to come, but also she taught me to be a better writer, line by line, as few others could.

I must also extend a deep-hearted thanks to my editor, Andrew Chadwick, who believed in this project from the get-go and committed to its long-term development at Oxford University Press. His invaluable advice and direction helped transform a rough early draft into the polished final product that you now hold in your hands. In addition, my acquisitions editor at Oxford University Press, Angela Chnapko, was incredibly helpful throughout the process, and I'm forever thankful for the time and energy she put into making it all happen.

While working on this book, I was fortunate to have the support of my amazing colleagues at the School of Communication and Media at Montclair State University. Thank you to Harry Haines, Hugh Curnutt, Christopher McKinley, Yi Luo, Todd Kelshaw, Marylou Naumoff, Christine Lemesianou, Merrill Brown, Larry Weiner, Vanessa Domine, Marc Rosenweig, David Sanders, Beverly Peterson, Kelly Whiteside, Tara George, Tom Franklin, Patricia Piroh, Tony Pemberton, Roberta Friedman, Susan Skoog, Lise Raven, Steve McCarthy, Stephanie Wood, and Janet Wilson.

If I've learned anything in course of my time in academia, it is that scholarship is only as good as its reviewers. In particular, I am indebted to the three readers at Oxford University Press who provided multiple rounds of feedback on drafts and drew my attention to countless issues, theories, and pieces of literature that I had previously overlooked or of which I had been simply unaware. I am pleased with how this book grew in size, scope, and sophistication throughout the review process, and this is largely an effect of the immensely thoughtful and in-depth suggestions they provided. I also owe major thanks to the anonymous readers at several other publishers who gave me valuable feedback during earlier stages of this project, as well as the blind peer reviewers at a long list of scholarly journals that considered (and sometimes even published) the research data that would ultimately find its way into this book.

A major goal of my work has been to relay the voices of citizens who engage in everyday acts of political expression online and offline, and this depends on the willingness of volunteers to participate in time-intensive interviews and share their thoughts and experiences in detail. In total, 101 people were interviewed for this project, and I am deeply grateful for the time and effort each of them put into the research process. Although their names have been changed to protect their confidentiality, I hope that their ideas and intellects shine through in the following pages.

Over the years, I have had the honor of engaging with and learning from countless outstanding scholars who have helped shape my imagination and my work. This (incomplete) list includes Katherine Sender, Michael Delli Carpini, John Jackson, Carolyn Marvin, Joseph Turow, Marwan Kraidy, Sharrona Pearl, Elihu Katz, Monroe Price, Klaus Krippendorf, Paul Messaris, Jeff Pooley, Mark Anthony Neal, Devon Powers, Melissa Aronczyk, Sarah Banet-Weiser, Emily West, Larry Gross, Caroline Dadas, Jennifer Stromer-Galley, Philip N. Howard, Lynn Spigel, Jeff Sconce, Lawrence W. Lichty, Hans Belting, and the late, great Richard Iton.

I also owe a big thank you to my fellow classmates at the Annenberg School for Communication at the University of Pennsylvania, for the friendship, the laughs, and the thought-provoking conversation. Just to name a few of many awesome people: Brooke Duffy, Brett Bumgarner, Michael Serazio, Mario Rodriguez, Lee Shaker, Matthew Carlson, Matthew Lapierre, Cabral Bigman, Heidi Khaled, Jeff Gottfried, Aymar Jean Christian, Bruce Hardy, Paul Falzone, C. Riley Snorton, Bill Herman, Khadijah White, Adrienne Shaw, Nicole Maurantonio, Emily Thorson, Brittany Griebling, Magdalena Wojcieszak, Seth Goldman, and Shawnika Hull.

In the course of developing my thinking for this project, I was inspired by a variety of artists who have explored the relationship between political expression and marketing aesthetics and forms in their work, including Ron English,

Barbara Kruger, Jenny Holzer, Shepard Fairey, Katharine Hamnett, Avram Finkelstein and ACT UP, and John Lennon and Yoko Ono. Thank you for making me think.

On a more personal note, I thank my fantastic friends, who have been there for me over these many years: James Hong, Peter and Brittany Micek, Michael Sherman, Marc Flury, Gilbert Galindo, Mark Siegmund, Elliott Ramos, Regina Yates, Katie Jefferis, Miles Levy, and Yaowei Yeo.

Finally, I owe everything that I've accomplished, including this book, to my wonderful family, especially Jacqueline and Dean, David and Adriana, Hilary and Fran, Fiona and Brynn, my Grandma Sue, and my late Bubby and Zayde, Flora and Marvin. I love you all and I thank you dearly.

The Citizen Marketer

1

The Citizen Marketer Approach
to Political Action

"Token" Symbolic Gestures?

When Krystal Lake arrived at her retail job one morning in early 2016 wearing a new hat, she wasn't expecting to cause a viral Internet sensation. The 22-year-old New Yorker simply wanted to make a political statement to shoppers in her local community who supported Donald Trump's campaign for U.S. president. Lake's hat, which she had imprinted with the words "America Was Never Great," was a parody of Trump's iconic "Make America Great Again" baseball cap that he wore at campaign events and sold to his supporters through his campaign website. Just as these Trump fans helped spread the campaign's marketing slogan to their fellow citizens by publicly displaying it on their foreheads—a material culture corollary to digital activities such as posting #MakeAmericaGreatAgain on social media—Lake took symbolic action to spread the idea that Trump was the wrong choice for the country. As she explained, "I kind of wanted to send a message . . . the message that other people are trying to send out is, like, America is like this wonderful place, like nothing bad ever happened . . . I feel like it was a lie." Lake, an African American, went on to cite the history of racial discrimination in the United States and remarked that the "point of the hat was to say America needs changing and improvement."[1]

What happened next came as a shock to her, although it followed a pattern that has become increasingly familiar in a political media landscape shaped by the peer-to-peer sharing of partisan digital content. A photograph of Lake wearing her hat, accompanied by outraged comments like "what an insult to #vets and ALL Americans," was posted on platforms such as Facebook and Twitter and quickly spread across networks of conservatives and Trump supporters.[2] To these people, the image was clear evidence that liberals against Trump were unpatriotic and anti-American, and by sharing the image on social media, they ostensibly promoted this idea further in the public discourse. Shortly thereafter, reports that Lake had received online death threats because of her hat's slogan—a stark indication of the feared power of political symbols—became a viral story of its own and was picked up by liberal news sites.³ Individuals who circulated news of the threats via links and posts on their social media feeds thus worked to raise awareness about the violent elements of the Trump campaign, presumably with the goal of helping turn the tide of public opinion against him. Back and forth the image of Lake and her hat traveled across the Internet, fueled by the partisan agendas of citizens who sought to use media participation as a way to impact the political world. Some were likely the sort of committed activists who also engage in activities such as fundraising for campaign organizations or door-to-door canvassing. Yet many others who shared the content on social media would be thought of as ordinary people—the sort who have little or no experience with traditional forms of organized political participation.

All in all, the saga of Krystal Lake and her contentious hat was another typical episode of political communication in the era of digital participatory culture. Yet although the sharing of viral images and articles on Facebook and Twitter may be one of the latest and most popular trends in political statement making at the level of the everyday citizen, it is far from the first. The fact that political slogans such as "Make America Great Again" and "America Was Never Great" are displayed on objects of clothing, as well as on digital platforms, underscores their connection to a tradition of symbolic political expression that dates back decades, even centuries. On the one hand, powerful public figures like Donald Trump have long had access to the media of mass communication to get their ideas across to the public—print newspapers and pamphlets, the broadcast airwaves, etc. On the other hand, nonelite citizens have historically lacked such access, yet they nonetheless have drawn on a range of personally accessible platforms of expression to enter the sphere of political communication and voice their perspectives in public. From the banners and sashes of the 19th century to the T-shirts and bumper stickers of the 20th, the act of publicly displaying one's political opinions and viewpoints with slogans, logos, and other symbolic markers has been a consistent feature of modern democratic life. However, as networked digital technologies have dramatically augmented the capacity of everyday people to participate in the circulation of media messages, political and otherwise, we are seeing an intensification of this kind of activity like never before. In previous generations, such expressive banner waving and sloganeering might have been dismissed as mere hoopla—a curious sideshow of popular democracy. In our contemporary era, however, it has become virtually impossible to ignore.

In fact, political media participation at the citizen level has become so pervasive in recent years that many have pointed to it as precipitating a crisis in how activism is defined in the 21st century. Often, it is the so-called digital native generation of young people that is blamed for transforming political activism into "slacktivism" or "clicktivism," among other pejorative terms that signal suspicion regarding the value of expressive media-based practices. The business writer Simon Sinek offers an exemplary critique, lamenting that, to their detriment, today's social media–savvy young generation "has confused real commitment with symbolic gestures."[4] In the scholarly literature regarding the Internet and politics, similar concerns abound. For instance, Evgeny Morozov argues that the private and personal nature of many digital media–based political actions tends to privilege comfort and self-satisfaction above all else. This "lazy," low-cost form of political activity, Morozov warns, may become a weak substitute for more intensive forms of political participation.[5] Stuart Shulman further suggests that when citizens see that their online political expressions are ineffective or simply ignored, they may be dispirited and become more cynical about participating in politics more broadly.[6]

As such critiques imply, the increasing centrality of symbolic forms of political activity has seemingly reached epidemic proportions in the age of social media, threatening to replace "real" activism with a more frivolous, less effective, and even disempowering mode of participation. The assumption that media-based political action may be insubstantial in and of itself is further embedded in the language that some researchers have adopted to describe it. For instance, a study led by Kirk Kristofferson seeks to investigate whether and under what circumstances an "initial token display of support" such as publicly joining a political Facebook group or wearing a political ribbon or button will lead to "more meaningful contributions to the cause."[7] The notion that media-based action can serve as a gateway to other forms of political participation, particularly at the organizational level, is important to consider—David Karpf argues that this idea is often what motivates political organizations to encourage their members to perform low-cost symbolic actions in the first place.[8] However, the use of the word "token" in this context reveals the essentially dismissive position that some analysts have taken toward expressive forms of political activity, both online and offline. As this line of thought goes, such activities may be a first step for citizens to do something meaningful in the political realm somewhere down the line, but on their own, they are superficial, shallow, powerless.

Yet is such a dismissal truly warranted? Does the simple act of posting a hashtag, link, or video, uploading a profile picture, throwing on a slogan hat or T-shirt, or slapping on a bumper sticker *do* anything politically, in its own right? Such a question has no easy answer because these tactics and their applications are constantly evolving. However, as this book will argue, the study of media-based symbolic action in the political realm can be significantly enriched by considering how it enables citizens to participate in political marketing, a field that has long used the circulation of symbols and media messages to promote political ideas and help shape political outcomes. In making this conceptual leap, we must take seriously the persuasive dimension of these expressive activities, which have rarely been given front-and-center attention in the scholarship. The explanation is perhaps obvious: at face value, it may seem preposterous to think that one's social media post or public display of a T-shirt or hat could influence others in a way that is politically consequential. Yet from the perspective of contemporary marketing practice, the notion of peer-to-peer influence as a networked, aggregate phenomenon is not only accepted wisdom, but also the very core of persuasive communication strategy in the digital age.

Indeed, one of the main reasons why it is important to consider political media-spreading activity within a word-of-mouth, or viral, marketing framework is that a wide range of organizational actors, from election campaigns to issue advocacy groups, have deliberately adopted this model as a way of extending the reach and credibility of their persuasive messages. For instance, in the context of U.S. presidential elections, Jennifer Stromer-Galley notes that campaigns are "using digital technologies to identify and empower supporters to persuade others in their own social network,"[9] an adaptation of Elihu Katz and Paul Lazarsfeld's two-step flow model of peer influence[10] that has inspired generations of word-of-mouth marketing initiatives. In other words, as supporters share a campaign's tweets, Facebook posts, and YouTube videos with their network of peers, they are not only expressing their own views, but also actively contributing promotional labor to the campaign as part of a carefully designed viral marketing plan. To offer an example outside of the electoral context (explored more in a later chapter), one of the most high-profile social media–based actions in recent U.S. political history, which involved millions of Facebook users displaying support for same-sex marriage by changing their profile pictures to a red equality symbol, was orchestrated by the professional marketing team of the large-scale issue advocacy organization Human Rights Campaign.

However, this kind of top-down effort to foster and direct peer-to-peer media promotion to expand organizational outreach represents only part of the terrain explored in this book. As a range of scholars have emphasized, much of the recent explosion of media-based political activity on the Internet has come not from traditional organizational entities at all, but from loose networks of like-minded individuals who converge online and engage in what W. Lance Bennett and Alexandra Segerberg call *connective action* around issues of common concern.[11] Although some aspects of this networked communication are more logistical in nature—including the use of online tools to coordinate offline actions such as protest demonstrations[12]—other forms of connective action center on the coordinated expression of political opinion in participatory media spaces. As Bennett and Segerberg note, networks of like-minded citizens often rely on "peer production and sharing based on personal expression" to cultivate public support for their causes and frame issues on their preferred terms.[13] Crucially, the authors identify this expressive dimension of connective action in terms of viral communication and the spreading of memes, defining the latter as a "symbolic packet that travels easily across large and diverse populations because it is easy to imitate, adapt personally, and share broadly with others."[14] The notion of promoting an idea by packaging it as a meme that can be shared from peer to peer is indeed the very essence of contemporary digital marketing logic. Although traditional political organizations have adopted these techniques more formally, taking cues from the commercial marketing sphere to harness online word of mouth for their promotion and advocacy efforts, the more informal application of the same techniques by grassroots citizen networks suggests that the concept of viral marketing has become imbedded in a much broader range of media-based political activism.

For instance, the #BringBackOurGirls hashtag campaign that became a runaway sensation on Twitter in 2014—part of an effort to raise global awareness about the abuses of the Nigerian militant group Boko Haram—was created not by a large advocacy organization with a hefty digital marketing budget, but by three Nigerian women who encouraged its spread at a grassroots level in hopes that "one voice can grow into millions eventually."[15] In addition, a wave of critical parodies of the #BringBackOurGirls campaign, which targeted president Barack Obama's foreign policy with slogans like #BringBackYourDrones and #WeCantBringBackOurDead, arose spontaneously among a fluid network of administration critics and left antiwar advocates[16]—an exemplary case of connective action of the expressive, viewpoint-spreading type. Essentially, such efforts call on citizens to act as microlevel agents in a networked circulation of ideas, disseminating symbolic packets of opinion and ideology as a means of influencing various sectors of the public. Whether orchestrated from the top down or emerging from the bottom up (or somewhere in between), these connective actions position their participants in a fundamentally promotional capacity: through liking, linking, sharing, and a range of other activities, citizens labor to extend and amplify the reach of favored political messages by passing them along and publicizing them to peers.

To be clear, the flow of persuasive political messages on the Internet and beyond is not restricted to this kind of peer-to-peer communication between ordinary citizens. Rather, it is situated within what Andrew Chadwick calls the hybrid media system, where pathways of message circulation often involve journalists, celebrities, and institutional actors in addition to members of the general public.[17] In the case of #BringBackOurGirls, for instance, the initial grassroots success of the hashtag on Twitter spurred the first lady, Michelle Obama, to lend her high-profile support in a post that was retweeted more than 70,000 times, garnering a wave of mainstream press attention that further contributed to its spread online.[18] In a similar vein, the hostile social media reaction to Lake's "America Was Never Great" image was greatly magnified by news coverage on right-wing websites such as *Breitbart*, and as noted, the subsequent online threats begat another round of headlines on left-wing sites like *Raw Story* that further added to the story's virality. Surely, professional journalists, political elites, and other major public figures play outsize roles in these hybrid media flows relative to ordinary citizens. The fact that it was Donald Trump himself who was responsible for launching the meme of "Make America Great Again" through a variety of channels, including his enormously popular official Twitter account,[19] handily illustrates that not all online influence is equal.

At the same time, however, the aggregate contributions of what Henry Jenkins, Sam Ford, and Joshua Green call grassroots intermediaries, defined as "unofficial parties who shape the flow of messages through their community,"[20] also deserve sustained attention. Indeed, data suggest that peer-to-peer media spreading can in fact be politically persuasive, at least for a portion of the population. According to a 2012 survey from the Pew Research Center's Internet and American Life Project, one-quarter of U.S. adults who use social networking sites claimed that they have become more active about a political issue after viewing posts or having discussions about it on these sites, and 16% claim that they have even changed their viewpoint on a political issue by engaging in such activities.[21] Moreover, although it may be difficult to quantify the magnitude of these sorts of effects on broader social and political outcomes, the widespread public participation in media-based symbolic actions such as hashtag and profile picture-changing campaigns suggests that millions of citizens *believe*—or at least hold some degree of hope—that such practices can make a meaningful difference in the world.

It is this broad set of participatory political promotion practices, along with the rationales that fuel them and the controversies that surround them, that make up the subject of this book. Its primary point of departure is to consider the long-term growth of these practices, magnified but by no means originated by the popularization of social media platforms such as Facebook and Twitter, as signaling an important shift in how political participation is conceptualized and performed in advanced capitalist democratic societies. The term *citizen marketer*—a play on the closely related concept of the citizen consumer—is an attempt to capture this emergent set of practices and corresponding logics. It is intended to give shape to a particular approach to political action that seems to become more and more high profile and widespread by the year—one that views participation in persuasive media-spreading activities as a privileged means of making a difference in public life.

To be clear, the term citizen marketer should not be taken to refer to an exclusive group that is solely defined by these kinds of practices. One of the primary goals of this book is to question the assumption that participation in symbolic actions such as posting political hashtags or memes necessarily replaces or substitutes for more traditional forms of political participation, as some slacktivism critics have warned. Rather than detract from real political action, these activities may provide citizens with an additional, complementary set of tools with which to support their interests and advance their political agendas. Furthermore, it may be helping to bring new entrants into the fold of political participation who may have otherwise remained inactive by opening culturally situated—and often affectively charged—spaces of engagement. As *The Citizen Marketer* argues, widespread participation in these sorts of activities may help invigorate popular democracy by casually injecting the political into the everyday spaces and places of popular culture. Scan your Facebook or Twitter feed, or even walk down a busy street or drive down a highway, and chances are that sooner or later you will see a citizen marketer flash a political message before your eyes that just might make you stop and think. Moreover, because of the participatory nature of these activities, there is great potential to democratize the field of persuasive political communication that has been historically dominated by elite interests and to elevate the voices and perspectives of marginalized groups.

However, another major goal of this book is to explore how the more controversial aspects of marketing and promotion practice introduce challenges that citizens must inevitably face as they adopt this approach to political action. Following Sarah Banet-Weiser's call to explore the ambivalent spaces of contemporary brand culture, politics included,[22] *The Citizen Marketer* brings into focus the trade-offs that come with expanding popular participation in politics by extending the marketing logics and languages that have long been used to secure the power of political, social, and economic elites. Marketing is indeed a dirty word in many circles, particularly among segments of the left that have identified neoliberalism and market capitalism as primary loci of contemporary political struggle. At the same time, some of these very critics—such as the *Adbusters* group led by the culture-jamming champion Kalle Lasn—have determined that the most effective way to push back against the forces of neoliberal capitalism is to co-opt its own marketing and advertising techniques to spread counterhegemonic ideas to the public.[23] This line of thinking will be explored in more detail in the following pages; my point here is simply to underline how the citizen marketer approach has become much broader than any one ideological constituency or bloc. Rather, it is being deployed by a wide range of citizens as a means of promoting and publicizing a variety of political ideas, including those that are broadly critical of elite uses of marketing in modern capitalist societies.

Following from this point, the tensions explored in this book often have less to do with the specific content of participatory political promotion campaigns—which can reflect both elite interests and counterhegemonic resistance, as well as both right-wing and left-wing ideological positions—than with their formal stylistic qualities, as well as their fundamentally declarative and side-taking nature. For instance, I examine how these practices are implicated in what numerous critics have denounced as the packaging of politics, raising important questions about the impact of marketing aesthetics on the evolving shape of contemporary political participation across ideological categories. Since the inception of modern political marketing more than a century ago by powerful elites, concerns have been raised about its capacity to trivialize complex issues by reducing politics to a simplified set of choices[24] and to manipulate the public through stylized and emotionally charged appeals that can conceal rather than reveal the substantive agendas of its propagators. When political marketing becomes more and more participatory, pulling a wide swath of citizens into its orbit and putting them to work as foot soldiers in the dissemination of packaged media persuasion, might these risks of trivialization and manipulation become exacerbated? Or, as Margaret Scammell argues in the context of election campaigning, does the shift to a more inclusive form of political marketing in an era of expanding media participation hold the promise of making it more authentic, more in tune with the concerns of ordinary citizens, and less in control of elites? Building on Scammell's call for a new political marketing literacy that can critically evaluate its contributions to democracy,[25] I argue in *The Citizen Marketer* that reconciling the inherent tensions of a packaged and marketed politics—that is, between style and substance, emotion and reason—is by no means impossible, but requires enhanced introspection and reflection on the part of all who now participate in its spread.

Furthermore, I explore how the citizen marketer approach intersects with the larger trends of political partisanship and polarization, which many critics have decried for potentially imperiling the quality of democratic life. As individuals are afforded more opportunities to take a side, show their support, and spread their preferred group's message through participatory media practices, the deliberative ideal of citizen-to-citizen interaction as a reciprocal exchange of ideas and a mutual forming of judgments may be significantly challenged. For some, we will see, expressive statement making may serve as a catalyst for two-way political dialogue. For others, however, it can elevate risks of social discord and ideological retrenchment. One core contention of this book is that the citizen marketer approach both closely aligns with and potentially strengthens an agonistic rather than deliberative model of democracy. The resulting emphasis on citizen media participation as a scene of group conflict rather than consensus building may be viewed as either perturbing or empowering, depending on one's normative and ideological commitments.

Neither a celebration nor a castigation of the trends it documents, *The Citizen Marketer* seeks to draw out the potential opportunities as well as the potential risks that are introduced by this broad shift in the conceptualization of political action. To help evince these subtleties, the book spotlights the voices of citizens who have participated in a range of political media–spreading activities both online and offline and interrogates their complex relationships toward the idea of promoting their political beliefs and persuading their peers. In contrast to some previous studies that report relatively little citizen interest in goals of peer persuasion when communicating about politics online,[26] the interview data presented in this book suggest how the citizen marketer approach is being enthusiastically embraced by some individuals who deliberately seek to exert political influence through their media-spreading activities. However, others who engage in the same practices describe more ambivalent feelings toward a viral marketing–like model of political action, reflecting the specter of unsavory manipulation that has loomed over political marketing practice throughout its long history. Still others reject this model outright in favor of alternative conceptualizations of civic communication practice, which, as we will explore, have complex and nuanced interrelationships with the citizen marketer approach (for a detailed description of the research, which encompasses three separate interview studies conducted over a four-year period and includes 101 respondents in total, please see the methodological appendix at the back of this book).

To be certain, a marketing-like, persuasion-centered approach to political action has not been universally adopted at the current juncture; moreover, it may only ever apply to a relatively limited, highly engaged subset of the citizenry who willfully contribute to what political scientists call "activated public opinion."[27] However, the goal of *The Citizen Marketer* is not to proffer an all-encompassing theory of citizen-level political communication in the digital age—surely a fool's errand given the complexity of the terrain. Rather, it is to bring to the surface a specific set of practices and logics that I argue has important consequences for the evolving shape of political participation and the character of democratic life.

Indeed, defining exactly what citizen marketing is and what it is not—where it begins and where it ends—is admittedly tricky because it can potentially overlap with a broad array of other citizen-to-citizen communication practices. In the following sections, I further delineate the citizen marketer concept by considering its relation to a range of relevant literature. As a first step, it is necessary to examine how a model of peer persuasion has been treated in the scholarship on citizen participation in political media discourse, and how this both diverges from and overlaps with other more widely discussed models that account for the political value of media participation. As we will see, there is an identifiable strand of literature that takes seriously the persuasive role of citizen-level political communication, although a conceptual connection to marketing practice has rarely been drawn. From there, I develop a more detailed argument about why a marketing framework is useful for making sense of the persuasive dimension of citizen media participation, and how it helps to illuminate both the risks and the opportunities that face those who seek to spread their political influence by participating in the circulation of media messages.

"A Battle over the Minds": Delineating the Persuasion Framework of Citizen Media Participation

Following the proliferation of networked digital media technologies from the 1990s onward, scholars have offered numerous ways to conceptualize the meaning and function of citizen-level participation in mediated political discourse. Two of these frameworks—broadly defined here as rationalist deliberative democracy and "culturalist" agency building—have been given extensive treatment in the literature. In contrast, the peer persuasion framework inherent in the citizen marketer approach has received far less systematic attention, although its lineage can be traced in the work of several key theorists, journalists, and activists. Indeed, all three of these frameworks are employed in various ways by the citizens interviewed for this book when explaining how they understand the meaning of their mediated acts of political opinion expression. Although *The Citizen Marketer* focuses primarily on this third framework, it also examines how a persuasion model overlaps in important ways with the two other major models that account for how citizen-level media participation takes on political significance, as well as how these models can sometimes come into friction with one another.

One of the most common approaches in the scholarship has been to examine how peer-to-peer mediated communication might provide forums for democratic deliberation and thus constitute a renewed public sphere. Influenced heavily by Jürgen Habermas's ideal of consensus building through rational and inclusive public debate,[28] the deliberative democracy framework emphasizes how citizens invested in the common good may use participatory spaces of interaction to produce "critically informed public opinion that can scrutinize and guide official decision-making processes."[29] In other words, the model privileges two-way exchanges of information and the rational debating of issues as a means of reaching decisions that institutional political actors can then follow. To be clear, this ideal of reasoned argumentation does include attempts to persuade the other party, yet Habermas stresses that both parties must be open "to the force of the better argument" and that "all motives other than that of the cooperative search for truth" must be eliminated.[30] We can thus draw a contrast between this deliberative framework and a model of persuasive communication motivated by a desire to simply spread one's deeply held viewpoints to others (which, as we will see, is often the case with regard to the kinds of activities considered in this book). Although the ideal of deliberative democracy has faced much criticism in recent years for its seeming elitism and its unflagging faith in Enlightenment reason,[31] it continues to be a popular way in which both scholars and citizens imagine the political value of participating in media discourse around political issues.

However, in contrast to a framework that emphasizes formalized decision making and consensus building, many scholars have drawn attention to more informal and culturally situated sites of citizen media participation—a move that seems to reflect the rise of popular networked digital platforms like Facebook and Twitter that combine political interaction with entertainment and peer socializing. This so-called culturalist turn[32] tends to focus on how politically themed interactions at the level of popular culture help build and strengthen collective identities and networks that can then be mobilized into action. According to Peter Dahlgren's theory of "civic cultures," the identities fostered through participation in informal peer-to-peer interaction—on social media and elsewhere—form the preconditions for more traditional forms of political participation by enhancing a sense of civic agency.[33] Maria Bakardjieva, whose work is also associated with this approach, argues that the moral and political identifications that emerge from everyday digital media participation form a "major reservoir of civic energy" that can potentially spur organizational participation at both the social movement and the institutional political levels.[34] Similarly, a group of researchers led by Jessica Vitak finds that young people's election-themed activities on Facebook help develop their political identities and foster civic skills that are necessary for future participation in the electoral political domain.[35]

In recent years, a substantial amount of research on politically oriented Internet use has lent empirical support to this broad body of theory connecting everyday online communication with various forms of collective political mobilization. For instance, Todd Graham, Daniel Jackson, and Scott Wright demonstrate how informal conversations about austerity policies among U.K. citizens in nonpolitical and culturally oriented online forums had a direct impact on subsequent organized actions, including protest demonstrations and coordinated campaigns to contact legislators.[36] Along similar lines, Jessica L. Beyer shows how discussion forums on popular culture–themed websites like 4chan and the Pirate Bay became a breeding ground for organizational mobilization around freedom-of-information issues and how this organized activity hinged on the group identities fostered by these online cultural spaces.[37] In addition, numerous large-scale survey studies have found a positive correlation between general social media use and offline forms of political participation, particularly in the context of joining social movement demonstrations and protest actions. However, in her meta-analysis of these studies, Shelley Boulianne points out that the evidence for causation is somewhat weak and that other variables, such as level of political interest, could explain the correlation.[38] Researchers have also found a positive relationship between social media use and offline civic activities such as joining group associations, which, according to social capital theory, is predictive of future levels of political participation.[39]

In sum, this broadly defined research framework understands the potential value of informal political interaction and expression on the Internet as an agency-building, identity-strengthening, preliminary step toward future political participation. Interestingly, several studies that seek to measure whether online communication will lead to offline political participation define the latter as including the symbolic expression of political viewpoints, albeit in material rather than digital form. For instance, Boulianne notes that the variable of political participation is often defined in a way that collapses together activities like voting and volunteering for election campaigns with decidedly symbolic actions such as displaying political buttons and signs.[40] This seems to suggest an assumption on the part of the researchers that this kind of expressive action *is* in fact a meaningful and valuable form of political participation—an endpoint as opposed to a stepping stone—although its precise import is largely unspecified. Adding to the confusion is the fact that some of the very online political practices that are seen as being identity- and agency-building precursors to real political participation, such as expressing opinions on social media platforms, are in a sense simply digital versions of offline symbolic activities like displaying political buttons or signs. Under such a schema, posting a political symbol on one's Facebook profile could be classified in an entirely different way than wearing that same symbol on one's lapel, although the motivation behind these actions—and perhaps even their impact on onlookers—may be similar.

To help clarify this complex set of issues, it is helpful to refer to Joakim Ekman and Erik Amna's detailed typology of political participation and civic engagement. The authors make an important distinction between *manifest* political participation, defined as "all actions directed towards influencing governmental decisions and political outcomes,"[41] and *latent* political participation, defined as prepolitical or "potentially political" forms of activity. In addition to civic association membership and community involvement (often defined in the literature as civic engagement), the latent category also includes the act of paying attention to political issues and having discussions about them with peers, both online and offline, as well as cultivating "a sense of belonging to a group or a collective with a distinct political profile or agenda."[42] Here, we can recognize the major themes of the above-outlined culturalist framework, in which peer interaction about politics builds identity and agency in a way that creates potential for future mobilization. By contrast, the manifest category, which represents what citizens *do* politically once they have been mobilized, includes communicative action intended to impact both the institutional and the extraparliamentary political spheres (along with a range of other actions that share this same basic set of goals, such as volunteering for election campaigns or joining social movement organizations and activist networks). Thus, for Ekman and Amna, "making one's voice heard" on issues of public concern is classified as manifest rather than latent participation because it is deliberately aimed at shaping political outcomes.[43] As they suggest, this can occur either by directly influencing public opinion, such as in the case of handing out leaflets on a street corner, or by applying pressure to decision makers, such as in the case of signing petitions. In both scenarios, we can identify persuasion as the defining characteristic of these expressive forms of manifest political participation.

Thus, according to Ekman and Amna's typology, communicating about politics online (as well as offline) could be considered either latent or manifest, depending on one's underlying motivations. The key distinction is the intended *instrumentality* of the expression, rather than any specific form or venue of communication. Of course, in real-world contexts of everyday media participation, the line between interacting about political issues as a noninstrumental means of political socialization and making one's voice heard on these same issues as a means of instrumentally shaping public opinion may be profoundly blurred. The interview research presented in this book attempts to draw out these nuances by exploring the often-overlapping and multifaceted motives that citizens assign to their political expressions on digital platforms and elsewhere. However, the value of drawing such a distinction here is that it helps avoid the pitfall of treating all forms of citizen participation in mediated political discourse as simply prepolitical steps to some other form of mobilization down the line. As I discuss in subsequent chapters, it is true that the kinds of expressive and symbolic activities examined in this book under the framework of marketing-like persuasion often contain a prepolitical, agency-building dimension, and this must be accounted for when assessing their broader political relevance. At the same time, however, we must not lose sight of the more instrumental and persuasion-oriented dimension of this expressive symbolic action, since it constitutes a significant component of the manifest participation that scholars point to as being the end goal of the latent stage.

As noted previously, some research focusing on the electoral context does include symbolic expressions of political opinion in definitions of manifest participation, although this is typically limited to offline activities such as displaying political buttons and signs (where a sense of "realness" is perhaps more immediately apparent). Yet the specific function of such expressive activities—including peer persuasion—does not warrant much in-depth attention in studies that treat participation in elections as a broad composite phenomenon. In contrast, a more robust framework of instrumental peer persuasion can be found in certain strands of scholarship that examine social movements and their uses of alternative media. Here, we can more clearly delineate a third approach to theorizing the value of citizen media participation, distinct from both the deliberative and the culturalist approach. For instance, Manuel Castells describes how new citizen movements that exist outside of the political mainstream use participatory media platforms to directly enter into contestations of power: "Because power relations are structured nowadays in a global network and played out in the realm of socialized communication, social movements . . . act on this global network structure and *enter the battle over the minds* by intervening in the global communication process" (italics mine).[44] Castells further posits that the Internet provides protest movements with a "means of acting on people's mind, and ultimately serves as their most potent political weapon."[45] In a somewhat similar vein, Kevin DeLuca and Jennifer Peeples argue that in the contemporary era of mediatized politics, activists strategically deploy mediated visual artifacts as so-called mind bombs that have the capacity to alter public opinion and, in turn, political events.[46] In these analyses, metaphors of warfare and weaponry (i.e. mind bombs, battle over the minds) are often employed to signal the instrumental force of these media-based symbolic actions.

As this vocabulary of mind warfare suggests, the peer persuasion framework of citizen media participation invokes a conception of political life that focuses on conflict, rather than on building consensus in the interest of the common good. In this way, it aligns with Chantal Mouffe's agonistic vision of democracy, which she offers as an explicit alternative to the deliberative democracy framework. As Mouffe argues, "what the deliberative democracy model is denying is the dimension of undecidability and the ineradicability of antagonism, which are constitutive of the political. By postulating the availability of a non-exclusive public sphere of deliberation where a rational consensus could be obtained, they negate the inherently conflictual nature of modern pluralism."[47] As a contrast, Mouffe's agonistic pluralism model emphasizes the centrality of contention in democratic processes, which would by definition include the sort of mediated battle over the minds described by theorists such as Castells.[48] In recent years, the theory of agonistic pluralism has been cited by numerous scholars as a more realistic way of describing today's polarized online political discourse than the high-minded notion of a rational deliberation of issues. For instance, Zizi Papacharissi argues that "Mouffe's emphasis on the agonistic foreshadows modes of political expression that have been popularized through the Internet," such as partisan blogs, activist YouTube videos, and contentious online political discussion threads.[49] Daniel Kreiss also invokes the agonistic model in his discussion of how the professional digital teams of election campaigns seek out supporters who will help them win media battles rather than question or debate their policy proposals, noting that "new media staffers generally focus on mobilizing preexisting selves who bring their ideological commitments to the public sphere."[50] In contrast to deliberative-minded citizens who solicit an open exchange of viewpoints in a cooperative search for truth, or citizens who communicate with peers about politics as a means of formulating their own identities and attachments, agonistic-minded citizens have already made up their minds, have already taken a stand, and use media participation as a means of fighting for their side.

In addition to being grounded in a fundamentally agonistic conceptualization of democracy, the notion of citizen activists marching into a media-centered battle over the minds to win political victories hinges on a broader set of assumptions about the primacy of sign and image in the political realm that broadly aligns with postmodern perspectives. For instance, Steven Best and Douglas Kellner argue that in a postmodern era in which "social life is filtered through the media, politics becomes a battle of images."[51] In fact, some theorists have posited that the overarching dominance of media images and signs in political processes—a set of developments sometimes described as *postmodern politics*—has shaped a generation of citizens who not only perceive political power to be equivalent to symbolic communicative power, but also are eager to get in on the action. Kevin Barnhurst, for instance, argues that in the minds of younger citizens raised in a postmodern, media-centered world, "the essence of political life . . . is the expression of opinions and preferences."[52] Similarly, John Gibbins and Bo Reimer claim in their book *The Politics of Postmodernity* that under conditions of media saturation, "the key method of political activity is voicing one's view and getting it heard. Politics in postmodernity is recognized to be constructed in language; politics is language"[53]—to which we might add all forms of symbolic expression, including images and video as well as words. What this suggests is that the broader mediatization of politics in the postmodern era has inspired citizens to act on the media—the realm of signs and images—in the hopes of making their own mark on the political world. As we will see, the self-aware and deliberate way in which many respondents featured in this book discuss goals of political persuasion when engaging in media participation suggests the intuitive appeal of this approach under postmodern conditions, as well as the need for scholars to understand this shift through a more robust theoretical lens.

To this point, perhaps the most explicit articulations of a media persuasion–centered approach to political action can be found not in the scholarly literature at all, but rather in two prescient texts of premillennial popular media criticism that are well steeped in a postmodern view of political reality: *Culture Jam* by Kalle Lasn and *Media Virus!* by Douglas Rushkoff. In *Culture Jam*, Lasn coins the term "meme warriors" to describe an emergent generation of citizen activists that privilege mediated public persuasion as a central political strategy: "Potent

memes can change minds, alter behavior, catalyze collective mindshifts and transform cultures, which is why meme warfare has become the geopolitical battle
of the information age. Whoever has the memes has the power."[54] Lasn further
lauds the Internet as "one of the most potent meme-replicating mediums ever
invented"[55] and calls on his fellow left-progressive activists to use digital networks
to spread ideas, influence minds, and ultimately affect global political change.
Rushkoff offers a similar vision in his treatment of viral media as an efficacious
political tool in a postmodern, media-centric age characterized by what he calls
the datasphere: "People who lack traditional political power but still seek to
influence the direction of our culture do so by infusing new ideas into this ever-
expanding datasphere. These information 'bombs' spread throughout the entire
information net in a matter of seconds."[56] Like Castells and others, Lasn and
Rushkoff draw on agonistic metaphors of warfare to suggest the instrumental—
and potentially powerful—impact of citizen media circulation on the shaping of
broader public opinion.

In addition to these martial allusions, metaphors of contagion in the writ-
ings of Lasn and Rushkoff—that is, the language of memes and virality that
have greatly come to define the era of networked digital media—signal how this
framework of citizen-level media participation aligns with contemporary digital
marketing logics. Both authors, whose 1990s-era works were written prior to
the proliferation of Web 2.0 and social media, arguably helped popularize the
notion that to effectively make a difference in a media-drenched world, one
must focus on injecting it with infectiously persuasive messages that can replicate and spread throughout the culture. Thus, Lasn writes that to "jump-start
this revolution . . . we just need an influential minority that smells the blood,
seizes the moment and pulls off a set of well-coordinated social marketing strategies."[57] Rushkoff's *Media Virus!*, although it avoids the explicit marketing-speak
of Lasn, nevertheless conjures a model of peer-to-peer influence that has proven
to be highly influential in the development of viral marketing more broadly since
its publication in 1994 (indeed, the book is often credited with inspiring the
viral marketing concept in the first place).[58] However, Rushkoff's primary focus
is on the use of viral media for political rather than commercial persuasion, as
he describes how the activists that he documents "depend on a worldview that
accepts that a tiny virus, launched creatively and distributed widely, can topple
systems of thought as established as organized religion and institutions as well
rooted as, say, the Republican Party or even the two-party system altogether."[59]

A key assumption underlying these biologically derived metaphors of contagion is that persuasive media messages act as uncontrollable forces, spreading from person to person through mere contact and ultimately overtaking the
public mind as a kind of ideological pandemic. Understandably, this perspective has been roundly criticized for proposing an overly mechanistic model of

peer influence. For instance, Henry Jenkins holds up Rushkoff's book as a prime example of "a flawed way to think about distributing content through informal or ad hoc networks."[60] For Jenkins, reducing people to "involuntary hosts of media viruses" undermines the role of human agency in the peer-to-peer spread of media messages, leading him to suggest that the term viral media be retired altogether in favor of the less deterministic "spreadable media."[61] Although *The Citizen Marketer* retains the viral metaphor as a useful, if flawed, concept for thinking about peer-to-peer political persuasion, it positions this issue of agency as one of central importance. After all, if those who circulate persuasive media messages imagine their efforts to be an instrumental means of political action, then they must necessarily see themselves as active agents of dissemination rather than passive conduits. Later in this chapter, I outline the concept of curatorial agency as a way of understanding how individuals might assert a degree of control over flows of media messages as they willfully contribute to networked and aggregate processes of making content "go viral" (to borrow a common cultural catchphrase).

The seemingly inescapable metaphor of viral influence also appears in Bennett and Segerberg's aforementioned discussion of connective action in digital networks,[62] as well as Papacharissi's work on affective publics as potential agents of political change.[63] Notably, however, Papacharissi offers an alternative conceptualization of cultural and ideological contagion that avoids the mechanistic overtones of biologism, evoking instead religious forms of practice. Drawing from Mary Douglas's anthropology of religion,[64] Papacharissi argues that collective expressions of political opinion on networked platforms like Twitter "may disrupt dominant narratives" in ways that resemble challenges to "orderly or cleaned structures of rituals."[65] As Papacharissi explains, "the potential for disruption and interruption derives from the fact that these narratives amplify visibility for viewpoints that were not as prevalent before."[66] This idea of citizens challenging the political status quo by making underrepresented perspectives and identities more publicly visible is a key theme that will be picked up in later chapters, as I explore how media-based symbolic action is imagined to work as a persuasive and effective form of political participation.

Papacharissi's model of religious-like contagion thus provides one compelling way of conceptualizing a peer persuasion–focused framework of citizen media participation. However, my core argument in this book is that our understanding of such a framework can be greatly enhanced by critically probing its relation to marketing logics and practices, particularly as they shift to more and more participatory and viral models in the digital age. To be sure, making this leap to a marketing framework is bound to make many uncomfortable, since the domain of marketing has itself been a frequent target of political and ideological critique. Those who are sympathetic to such a viewpoint may therefore be tempted to avoid the connection between viral marketing and viral politics altogether and seek out a framework of peer persuasion that is wholly independent of marketing discourses. As should be clear by now, my approach in this book is quite the opposite. However, rather than simply pushing these critiques of marketing and its potentially worrisome effects on society to the side, I contend that they help bring to the surface many of the key issues that loom over a media persuasion–centered model of political participation more broadly, regardless of whatever viewpoint or ideology is being made to go viral. In the following section, I examine this body of criticism as a way of highlighting what is at stake in the hybridization of politics and marketing logics. Indeed, the citizen marketer approach is by its very nature embroiled in these tensions, even as it promises to democratize the domain of marketing that has long been an instrument of power and influence for political and economic elites.

The Citizen Marketer in Context: Promotional Culture, Packaged Politics, and the "Marketplace of Ideas"

To better understand the tensions and fault lines surrounding the citizen marketer approach, it is necessary to begin by considering the broader ascendancy of marketing logics and models in the political cultures of advanced capitalist democracies. For Andrew Wernick, the 20th century was largely characterized by the increasing dominance of promotional practice in all areas of these societies, politics included. Wernick introduces the term *promotional culture* to refer to the way in which promotional messages have become "co-extensive with our symbolic produced world."[67] What this means is that in consumer-based economies that depend on the promotional industries for their very functioning (advertising, public relations, etc.), marketing becomes a kind of uberlogic that comes to pervade all social activities and institutions—even those that exist outside of the commercial marketplace. For Wernick, this promotional culture has fully penetrated the domain of institutional politics, resulting in the explosion of slick election campaign advertising and the branding of politicians. Under conditions of an all-encompassing promotional culture, the practice of politics itself becomes largely a matter of circulating competing promotional messages, particularly via technologies of symbolic mediation. In line with numerous theorists noted previously, Wernick associates this shift with the predominance of the image and the sign in postmodernity.

In the United States, in many ways the global leader of institutional political spectacle from the 19th century onward, the promotion of electoral candidates has been deeply intertwined with commercial marketing practices.

The historian Liz Cohen traces the use of consumer-oriented mass marketing techniques within U.S. elections to as early as the 1890s, noting how their use intensified starting in the 1930s when both the Democratic and the Republican parties began hiring advertising professionals to promote their candidates. By the 1950s-era presidential campaigns of the Republican Dwight Eisenhower, many of New York's top advertising agencies, such as BBDO and Young & Rubicam, were on board, making electoral marketing nearly indistinguishable from the pitching of household products. Over the years, numerous observers of the U.S. political system have been highly critical of this long-term move toward a mediatized, marketed, and merchandized form of politics. Back in the 1950s, fresh from his defeat at the hands of Dwight Eisenhower and his well-remembered "I Like Ike" advertising campaign, the former U.S. presidential candidate Adlai Stevenson remarked that "the idea that you can merchandise candidates for high office like breakfast cereal . . . is the ultimate indignity to the democratic process."[68] Similar critiques of the use of corporate-style marketing techniques in presidential campaigns can be found in political commentary throughout the postwar period, from Joe McGinniss's *The Selling of the President, 1968*[69] to Kathleen Hall Jamieson's Reagan-era *Packaging the Presidency.*[70] As the United States exported this model of candidate marketing across the world from the 20th century onward, many critics have similarly lambasted it for its seemingly problematic elevation of political style over substance. For instance, the Australian scholar Elaine Thompson argues that "television, advertising, polling and image making have all been transmitted from America to Australia and have helped change the nature of election campaigns . . . these changes have helped trivialize issues."[71] The common concern among these critics is that as politics becomes packaged and marketed more and more like a commercial product for mass consumption through television and other mass media, superficial concerns will inevitably trump substantive ones and leave the democratic process impoverished in the process.

Crucially, these worries over political trivialization are closely linked to concerns about undue public manipulation by elites. As John Street explains, the critique of packaging in politics typically refers to "some form of deception: [packaging] is a device for distracting attention from the content, of presenting things in a way that fails to reflect their true character." The feared result of packaged politics, Street notes, is a "political system in which cynical politicians and their party managers seduce and delude the voters," essentially tricking them into supporting candidates and policies that they wouldn't have otherwise had they only been presented with more accurate and truthful representations.[72] This parallels well-established critiques of commercial marketing, where the concern is that style-heavy advertising campaigns deceive consumers into buying products that they don't really need by associating them with desirable social values that aren't really there, and by deliberately withholding useful product information in favor of appeals to fantasy and emotion that are designed to cloud one's better judgment.[73]

Just as critics warn of the power of commercial advertising to expand corporate power through deception and seduction, critics of political packaging warn of how institutional political elites may use the same formal techniques to manipulate and extend their control over the public. In response to these arguments, Street mounts a defense of political packaging as a more wide-ranging and politically diverse adaptation to postmodern media culture that does not necessarily divorce form from content and style from substance. Moreover, Street argues, packaging may actually strengthen democracy by making politics more accessible to ordinary citizens. Similarly, Scammell posits that rather than being an exogenous force that has invaded the political realm, marketing logics are "inherent in the structure of political competition," and furthermore, the populist aesthetics of marketing and advertising open important spaces for the experience of emotion and pleasure in democratic processes.[74] These compelling counterarguments will be noted and engaged with throughout this book. However, it is important here to establish what is at stake in the packaging process—that is, the interconnected perils of trivialization and manipulation—at a time when political marketing is becoming more and more a site of widespread citizen participation.

When addressing the potentially manipulative dimensions of political marketing, one must also inevitably contend with its historical connection to the even more maligned and feared practice of propaganda. Although definitions of political propaganda have ranged widely over time, the word is now typically invoked to describe the all-encompassing persuasion efforts of totalitarian regimes and military and religious powers rather than with the political communication systems of modern democracies. Despite attempts to frame it in more neutral terms as simply any sort of political persuasion in symbolic and mediated form, propaganda has not been able to shake its pejorative meanings stemming from its connotations of deception and trickery.[75] Indeed, Nicholas O'Shaughnessy describes propaganda as "a form of coercion without the appearance of coercion," underscoring its associations with misleading, disingenuous, and underhanded practices of political persuasion. Furthermore, O'Shaughnessy notes the historical link between modern advances in propaganda (which includes, in his definition, election campaign television advertising, particularly of the negative fear-mongering type) and the techniques of commercial marketing, noting that in the 20th century "there was a transmission of learning from the world of commerce: a market culture rests upon an edifice of persuasion and it is not surprising that political activity should absorb these methodologies."[76] In fact, this exchange went in both directions, as epitomized by the career of Edward Bernays, a 20th-century master of media persuasion who traversed the fields of business and politics. Bernays famously pioneered the use of Freudian psychology to manipulate publics by appealing to the irrational impulses of the unconscious, applying his techniques to both government propaganda and commercial public relations (the latter being a term he himself coined as a more positive-sounding substitute for propaganda—a perfect encapsulation of his handicraft).[77] As the legacy of Bernays reminds us, the histories of propaganda, commercial marketing, and modern political advertising and promotion are all bound up with one another, connected by a thread of potentially deceitful manipulation that continues to haunt the terrain of political persuasion in the contemporary era.

If political elites have been transformed by an all-encompassing promotional culture into manipulative, even propagandistic media marketers, then where does this leave ordinary citizens? The most pessimistic view, articulated by Frankfurt School–influenced theorists such as Stuart Ewen, suggests that the promotional media spectacle of elite institutional politics and its "dominance of surface over substance" ultimately has a pacifying and disengaging effect on citizens—a sort of new opiate of the masses.[78] However, this idea has been challenged in recent years by cultural studies scholars who emphasize the agency of the political audience to interpret and actively construct meaning out of the persuasive marketing messages handed down from above by political elites.[79] Writing off the public as a bunch of brainwashed dupes, so they would say, ignores the possibility that many media-literate citizens are becoming savvy about how the system works. At the same time, Barrie Axford notes that although interpretative agency can help citizens resist elite political manipulation and thus potentially assuage the sorts of fears noted above, this alone "can still leave them without the means and perhaps the desire to affect the process" of a politics marketed from above.[80]

This concern over the relatively weak role of citizens in political systems dominated by elite media marketing and promotion ties in with broader critiques of the marketplace of ideas as a model of democracy based on (neo)liberal economic principles. Associated with mid-20th-century thinkers like Anthony Downs and Joseph Schumpeter, this model envisions a system in which politicians competitively sell themselves and their policies to voters in a way that mimics commercial markets, and has provided the conceptual foundation for modern political marketing as a professionalized practice.[81] For many critics, the underlying problem with the marketplace-of-ideas model—above and beyond any concerns over the manipulative tactics that may be employed in the selling process—is that it positions citizens in a consumer-like role that is seen as fundamentally passive and limiting. For instance, Justin Lewis, Sanna Inthorn, and Karin Wahl-Jorgensen argue that in advanced Western capitalist democracies steeped in neoliberal values, "the world of politics is left to the politicians and the experts, who compete to sell us positive images of themselves . . . [consumers] respond to the possibilities on display rather than setting the agenda . . . their power is limited to the ability to choose one product rather than another . . . as a way of being, it falls far short of the ideal of citizenship that underpins democratic society."[82] Echoing these concerns, Nico Carpentier argues that the liberal marketplace-of-ideas model carves out a minimalist role for democratic participation that is limited to voting-as-buying, resulting in a weakened democracy dominated by elite rather than citizen control.[83]

Furthermore, critics have raised concerns about how a consumer-like model of democratic citizenship emphasizes the pursuit of individual self-interest over a more civic-spirited concern for what is best for the larger community. For instance, Liz Cohen discusses how the system of modern institutional political marketing has mimicked the commercial marketing sphere in emphasizing audience segmentation and highly tailored appeals, leading to the troubling trend of "purchaser citizens" who act like self-interested consumers in their interactions with government. Instead of asking how various policies might benefit the greater good, Cohen argues, purchaser citizens, shaped by their experience of consumer-tailored political messaging, ask only how these policies might benefit them personally.[84] Along similar lines, Heather Savigny argues that a marketer-consumer model undercuts democracy because of its "inherent individualization of both the public and political actors, and emphasis upon their differing short-term goals," which she connects to the troubling encroachment of neoliberal ideology more broadly.[85] Although Asard and Bennett advocate for a more neutral and context-contingent understanding of the marketplace-of-ideas concept, their analysis of modern political marketing trends leads them to warn of "a possible eclipse of traditional political cultures through the continual initiatives of marketing research and strategic communication. Messages that target individuals at deep emotional levels may tear down the collective associations that draw citizens into public life."[86] In other words, as citizens are increasingly addressed by political marketing professionals in terms of their personal needs and desires as consumers, they may lose sight of the broader public interest and foreclose their own potential to contribute to collective and civic goals.

Thus, the consumer-like citizen who emerges from a liberal marketplace-of-ideas model of democracy is envisioned by critics to be self-interested, politically delimited, and ultimately demobilized. A similar picture emerges from scholarship that examines the effects of corporate encroachment into spaces of political activism that lie outside the institutional or parliamentary sphere. For instance, Alison Hearn argues that so-called corporate social responsibility efforts like Disney's pro-environmental online venture advance an individualistic, self-centered, consumption-celebrating style of activism that diminishes the kind of collective consciousness necessary to organize and effect meaningful political change.[87] Along similar lines, Dahlgren laments the rise of consumption-centered activism because of its seemingly inextricable ties to a neoliberal model of "intensive commercialization and values that increasingly affirm private fulfillment over social solidarity."[88] Interestingly, Dahlgren connects this idea to the earlier-noted critique of slacktivism, describing personalized digital expression as an essentially consumerist form of political participation "where feeling good takes priority over political commitment."[89] For Dahlgren, the feared result of "societal participation via commercial logics" that privilege personal pleasure and satisfaction is a "weakened political efficacy"[90]—again, a key theme that cuts across a range of critical scholarship on the hybridization of democracy and market relations.

However, the pessimistic conclusion that consumption activity—and marketing activity, for that matter—is fundamentally at odds with meaningful democratic participation is by no means universally held. In the next section, I examine how the concept of the citizen consumer has been used to advance a more optimistic vision of "politics by other means" and how the concept of the citizen marketer emerges from this line of thought and potentially continues this recuperative project. In contrast to scholarship that positions the roles of citizen and consumer as fundamentally antithetical to one another, the citizen consumer literature alerts us to the complex ways in which these roles might potentially be reconciled, creating new and alternative routes to political participation and empowerment. At the same time, the litany of criticisms of the market–democracy nexus outlined above—from leaving citizens vulnerable to trivialization and manipulation to prioritizing self-interest (or perhaps factional group interest) at the expense of the broader public interest—help sensitize us to the profound issues that must be navigated as the domains of both consumption and media promotion are increasingly embraced as sites of political action.

From the Citizen Consumer to the Citizen Marketer

To properly elucidate the concept of the citizen marketer, it is necessary to examine how it extends from the logic of the citizen consumer. Broadly speaking, the citizen consumer can be understood as a person who consciously draws on his or her status as a citizen when making consumption choices in the commercial marketplace or, as Cohen describes, "put[s] the market power of the consumer to work politically."[91] In its most typical usage, the citizen consumer describes an orientation of individual social actors who take into consideration the political repercussions of their consumption behaviors and make deliberate choices about whether to purchase specific products in accordance with their political interests. Using the allocation of their shopping budgets as a political tool to support certain kinds of favored businesses, industries, and production practices—as well as to deny financial support to others that they perceive as problematic—citizen consumers work to reshape the marketplace of consumption into a venue of political participation. In her historical analysis, Cohen emphasizes a distinction between this model of the citizen consumer and the more self-interested purchaser citizen noted previously; whereas the latter acts like a consumer in the realm of politics, the former acts like a citizen in the realm of the marketplace. In other words, rather than prioritizing personal pleasure and satisfaction, citizen consumers are seemingly guided by altogether more public and civic concerns, even as they assert their political will through individual acts of consumption.

This politicization of personal consumer activity has been identified by scholars as part of a broader shift in advanced capitalist democracies toward what Bennett calls "lifestyle politics," which correlates with a long-term decline in citizen participation in traditional political institutions. As Bennett explains, the "dutiful" citizen of the past felt an obligation to participate at this institutional level and saw activities such as voting and formal party membership as the center of political life. However, because of a range of factors, including increasing public cynicism toward political institutions, as well as the growth of consumer culture and its attending values of individual self-expression and empowerment, today's citizens often look outside traditional political institutions and organizational collectives to advance their political agendas.[92] These "actualizing" citizens, Bennett argues, are guided by a "higher sense of individual purpose,"[93] which may or may not correspond with their own material self-interest, but certainly aligns with their closely held ideological commitments and attachments. Addressing the Latin American context, Nestor Garcia-Canclini posits that the appropriation of consumption for new forms of activism centered on issues such as environmentalism, race, and gender is a development rife with democratic potential, as corrupted political institutions are effectively sidestepped in favor of more directly accessible sites of participation. As Canclini argues, "questions specific to citizenship . . . are answered more often than not through private consumption of commodities . . . than through the abstract rules of democracy or through participation in discredited political organizations."[94]

This more optimistic assessment of the citizen consumer as a form of "politics by other means" is also reflected in Bennett's work on what he terms "logo activism," which marks a key turning point in the scholarship. As Bennett argues, the politicization of consumer culture now extends beyond mere consumption activity (e.g. boycotts and "buycotts") to include activist efforts to appropriate the symbolism of brands to launch political critiques of the transnational corporate entities that deploy them for economic gain. Specifically, Bennett examines networked digital communication campaigns, largely grassroots in nature, that co-opt and parody brand logos like the Nike "swoosh" to draw widespread attention to the human rights, labor, and environmental issues that surround these very corporate actors. The advantage of logo activism, Bennett posits, is that it repackages radical political ideas in a branded language that is instantly recognizable—and thus potentially more accessible and impactful—to a consumer-oriented public that may be less responsive to older, drier forms of political advocacy.[95]

This notion of using the stylized appeal of marketing techniques to critique the marketers, so to speak, represents the core logic of culture jamming, a thoroughly postmodernist approach to activism that seeks to "destabilize and challenge the dominant messages of multinational corporations and consumer capitalism" by directly hijacking and subverting them.[96] Although its conceptual lineage has been traced to the Situationist International movement of the 1960s and related figures such as Guy Debord, the term was first coined by the experimental music group Negativland in the 1980s in reference to public billboard alteration and other radically appropriative political art projects.[97] It was then further popularized in the 1990s by *Adbusters* magazine, known for its anticonsumerist parodies of commercial advertisements. Crucially, the development of culture jamming and logo activism points to a politicization of marketing activity that parallels the politicization of consumption activity denoted in the concept of the citizen consumer: the latter seeks to co-opt the domain of commerce as a venue for asserting one's political will, and the former seeks to co-opt the closely related domain of commercial promotion and media advertising for similar ends.

Furthermore, Jamie Warner notes how culture-jamming techniques of marketing parody and subversion are being applied to the domain of institutional politics, in addition to the typical corporate targets of logo activism and outfits like *Adbusters*. Warner explains that "as politicians and political parties increasingly utilize the branding techniques of commercial marketers to 'sell' their political agendas, it follows that similar jamming techniques could be employed to call those branding techniques into question."[98] For instance, the "America Was Never Great" hat, created by Krystal Lake as a critical appropriation of the Trump campaign's marketing efforts, serves as a vivid illustration of what political culture jamming can look like in the contemporary context.

Examining the logic of culture jamming draws our attention to how the idea of political marketing has transcended its origins as an elite practice and has been adopted by a wide range of constituencies as a means of disseminating political ideas in the postmodern language of promotional culture. Now that both hegemonic and counterhegemonic forces are exploiting the formal techniques of media packaging to advance their positions in the public arena, citizens from across the political spectrum have opportunities to participate in the circulation of political marketing—or political culture jamming, if they prefer—as they spread their favored ideologically charged content to their peers. In fact, as addressed above, Lasn's *Culture Jam* can perhaps be considered the most clearly articulated manifesto for a marketing-centered approach to political action more broadly.

However, the phenomenon of culture jamming alone does not fully encapsulate the citizen marketer concept as advanced in this book. This is because the discourse of culture jamming tends to focus on the production of marketing-like political media content, rather than on its circulation. *Adbusters*, for instance, is a professionally produced magazine and does not in itself involve any participation on the part of ordinary citizens. The creation of this kind of alternative political content may significantly expand the range of ideas that citizen marketers can disseminate through their own media-spreading activities, but it should not be confused with the promotional labor of citizen marketing itself.

To distinguish exactly what domain of social activity is being seized as a venue for political action in the citizen marketer approach, it is necessary to examine how the consumer–marketer relationship has evolved in the era of social media and peer-to-peer networked communication. Although the concept of word-of-mouth promotion has existed for many decades and predates the Internet, the traditional marketing model has tended to focus on a one-way communication process between active agents of persuasion and passive audiences of consumers.[99] However, in recent times, that has all changed dramatically. With the rise of networked digital media, ordinary people are increasingly engaging in micro-level acts of promotion, whether they fully understand and appreciate it or not. Every time an individual posts a link to a movie trailer or music video, lists a favorite television show on a personal profile, presses the "like" button for a commercial brand (automatically triggering posts to one's connections), or retweets a celebrity update, he or she is engaging in promotional labor, often described in the industry as electronic word of mouth.

As Banet-Weiser explains, the marketing strategy undergirding contemporary social media platforms "engages the labor of consumers so that there is not a clear demarcation between marketer and consumer, between buyer and seller. The engagement of consumers as part of building brands . . . potentially engenders new relationships between the buyer and the bought."[100] As noted earlier, Jenkins, Ford, and Green use the term grassroots intermediaries to refer to this concept of media nonprofessionals coproducing a brand by voluntarily spreading its promotional messages to peers. Through these practices, grassroots intermediaries "may become strong advocates for brands or franchises. Grassroots intermediaries may often serve the needs of content creators, demonstrating how audiences become part of the logic of the marketplace."[101] Indeed, the promotional industries of the early 21st century have become greatly focused on fostering and encouraging this kind of participatory marketing activity, coaxing social media users to share an abundance of branded viral content in hopes that this peer-to-peer communication will be more persuasive than traditional paid advertising (which has shown to be far less trusted by the public).[102] In the industry literature, this emergent consumer-as-marketer goes by many names, including the brand advocate[103] and the customer or brand evangelist.[104]

In addition, the term citizen marketer has also been used in commercial digital marketing contexts, although it is meant in a different way than how I define it in this book. For instance, Jackie Huba and Ben McConnell describe citizen marketers as ardent fans who use Internet tools to create amateur promotional messages on behalf of their favorite brands, products, and people: "they are on the fringes, driven by passion, creativity, and a sense of duty. Like a concerned citizen."[105] In addition, Ruth E. Brown describes citizen marketing as the act of "consumers voluntarily posting online product information"[106] through reviews, videos, and other forms of user-generated content that can influence the purchasing decisions of their networked peers. In both cases, the authors equate citizen marketing with the amateur production of promotional messages, which, again, is not how I deploy the term. As in the case of culture jamming, this sort of amateur production activity may expand the available range of persuasive media content that can be spread peer to peer, beyond that which is created by professionals and traditional elites. However, thinking of citizen marketing only in terms of amateur production obscures the promotional labor that people engage in as they publicize favored content through practices of circulation, sharing, and endorsement.

Furthermore, the word "citizen" is invoked in these commercially oriented treatments of the citizen marketer more as a synonym for an ordinary person or media amateur than as a way of signaling any distinctly political activity. At the same time, however, Brown's emphasis on how networked consumers are driven to share both positive and negative assessments of commercial brands—a point also made by Jenkins, Ford, and Green in their discussion of grassroots intermediaries as having the potential to be brand hostile as well as brand supportive[107]—suggests how consumers may assert a degree of power over the commercial marketing process as they take advantage of their own peer influence on digital platforms. To be certain, this kind of industry-focused literature is intended to help professional marketers figure out how to manage and shape electronic word of mouth for their own strategic purposes. Yet it is also acutely aware of the struggle to adapt to a more participatory media environment in which consumers have in some sense democratized the flow of persuasive messages. In this sense, Brown's take on the citizen role of what she calls citizen marketing in the commercial context dovetails with a key theme of this book: as the field of media promotion becomes more and more a site of public participation in the digital age, there is potential for a broadening of voices and perspectives and an increased opportunity for grassroots actors to challenge the power of the institutional elites that have traditionally controlled the machinery of persuasive communication.

Although the role of the grassroots intermediary (i.e., the brand advocate or evangelist, etc.) has largely been examined up to this point in relation to the commercial sphere, this book is concerned with how it becomes a site of participation in the political sphere, in the broadest of possible senses. In other words, my focus is on how people use their promotional power as microlevel, peer-to-peer agents in the broader information environment to help advance a political—as opposed to a purely commercial—agenda. In some cases, like the logo activism campaigns discussed previously, this may in fact involve targeting the corporate sector on distinctly ideological grounds. However, unlike earlier definitions that are limited to the domain of the commercial marketplace, my definition of the citizen marketer refers to peer-to-peer media-spreading activity that is motivated by an interest in promoting one's political opinions and agendas to one's peers. In other words, the citizen marketer consciously draws on his or her status *as a citizen* when making choices about participating in the circulation of mediated symbolic artifacts, a practice that by its very nature affords free promotion and publicity to various ideas, organizations, and people. In the era of networked digital media and the social sharing of content, this promotional labor has become as routine as going shopping at the supermarket, so perhaps it is not surprising that it has been widely embraced as a venue for politics by other means in a way that parallels the citizen consumer model.

Furthermore, as mentioned earlier, this kind of citizen marketer activity is now being deliberately fostered by a wide range of political institutions and organizations, which are following the lead of the commercial sphere in attempting to exploit the persuasive power of electronic word of mouth for strategic ends. Regarding electoral campaign promotion on social networking sites like Facebook, for example, Michael Serazio explains that "as part of wider ambitions by the advertising industry to colonize these spaces, [political] consultants are eagerly pursuing the recruitment of evangelists there, which represents . . . 'the holy grail for campaigns.'"[108] As this language underlines, the political equivalents of brand evangelists are not wholesale inventions of the professional class of digital marketing consultants, but emerge more organically from the broader technological and cultural conditions of the social media age, to the point where their promotional labor can be "recruited" by institutional actors. Thus, examining the citizen marketer approach only in the context of formalized, institutionally architected viral promotion risks missing the full scope of the phenomenon, which spans both grassroots connective action and participation in professional political marketing campaigns.

However, the story of the citizen marketer does not begin with Facebook and Twitter, or with the rise of contemporary viral and social media marketing strategies, although these developments have made this approach to political action far more widespread and visible. If we expand our understanding of media content to include artifacts inscribed with text, graphics, and other symbolic material, we can begin to put the long and complex lineage of citizen marketing into focus. For instance, for those who wish to spread a persuasive political message to a public audience but are without access to traditional media outlets (i.e., newspapers, radio, television, etc.), the corporeal human body has historically been mobilized as a kind of ad hoc exhibition screen. As Stuart Hall once noted with regard to black popular culture, "these cultures have used the body as if it was, and often it was, the only cultural capital we had,"[109] underlining how this personal space of signification has been utilized by even the most historically marginalized and disempowered groups. Indeed, printed buttons, T-shirts, caps, handheld signs, and banners have been used by generations of citizens from across the spectrum to inject political messages into public space. In addition, a range of conspicuous personal possessions have similarly been transformed into media screens for political messaging purposes, most notably the automobile (i.e., bumper stickers) and the home dwelling (i.e., lawn signs). Of course, these word-of-mouth—or word-of-chest, word-of-car, etc.—promotion techniques have long been exploited by elites for institutional political marketing, just as today's digital consultants are currently working to take advantage of content sharing and endorsement on social media platforms like Facebook and Twitter.

This history of politically themed, message-spreading consumer goods underscores how the citizen marketer has grown directly from the citizen consumer: in the case of slogan T-shirts or bumper stickers, citizens not only use their politics to guide their consumption choices at the cash register, but also display the purchased goods as a means of publicizing the viewpoints expressed therein. A parallel can be drawn to the act of sharing political media content on digital networks: at once, an individual consumes the content that she or he favors and conspicuously displays the consumption choice in ways that expand its publicity (through liking, linking, reposting, etc.). Thus, relatively primitive personal platforms for exhibiting media material, such as T-shirts and bumper stickers—which are limited to displaying static, two-dimensional graphics and short bits of text—can be understood as forerunners to today's social media profile pictures, hashtags, and status updates. Although each of these platforms has its unique technological and expressive capabilities, what they have in common is that they all provide media nonprofessionals with microlevel (but not necessarily insignificant) opportunities to participate in the spread of political messages in the service of advancing their interests. It is no mere coincidence, then, that slogans like "Make America Great Again" and "America Was Never Great" would appear both in digital posts and on displayable objects like baseball caps, since both types of venues can be used as sites of conspicuous symbolic consumption to make one's politics visible to a public audience.

One of the primary contributions of *The Citizen Marketer* is thus to draw connections between these forms of participatory and embodied media practice that may not have previously been recognized as related. My argument, however, is not simply another version of "there is nothing new under the sun"— that is, that networked digital media simply offer new ways of doing the same old things. In fact, much of this book is dedicated to exploring how digital technologies are both intensifying and transforming the citizen marketer approach in many crucial ways, particularly in terms of allowing citizens to strategically share journalistic information with peers as well as more concise and packaged symbolic expressions. Furthermore, I do not mean to suggest that older artifacts like placards and buttons indicate an age-old tradition of participatory political marketing that has been with us since the dawn of time. Although important antecedents do stretch far back, the citizen marketer approach has developed in response to specific political, economic, and cultural conditions—a point that will be addressed in detail in the following chapter, which explores the citizen marketer's historical lineage.

Selective Forwarding and Curatorial Agency

In making the connection between digital and material culture forms of political message spreading, this book focuses on a specific aspect of participatory media practice that demands sustained attention: what I refer to as *selective forwarding*. By this, I mean the act of selecting preexisting media material that one favors and sharing it with others via a platform that she or he can personally program or control—essentially a matter of media distribution, rather than production. Whereas the creative political expressions of citizens working in mediated contexts—blog authors, YouTube video producers, Internet meme makers, etc.—have been the subject of much scholarly interest and discussion in recent years, the practice of passing along the creative expressions of others to a public audience is only now beginning to emerge as a major focal point of theory and research. This is perhaps because original creative work may be seen as innately more interesting and exciting than "mere" relay, which, by contrast, can appear mundane, rote, or facile in character. In discussions of human agency, political and otherwise, it is far easier to recognize how the act of crafting unique creative expressions provides individuals with a potentially influential voice. However, according to the viral model that has come to pervade much of participatory media culture, it is the peer-to-peer *circulation* of information—as opposed to its initial moment of creation—that truly marks its salience and resonance among the public. Thus, the act of nodal distribution takes on crucial significance in a framework of viral marketing–like political persuasion, as each individual agent of dissemination contributes to the overall spread of a message.

As noted earlier, critics of viral metaphors of contagion have rightly emphasized the human agency involved in the spread of media material, although they typically point to creative processes of transformation as their primary evidence. For instance, Limor Shifman writes that "in the digital age, people . . . can spread content *as is* by forwarding, linking, or copying. Yet a quick look at any Web. 2.0 application would reveal that people do choose to create their own versions of Internet memes, in startling volumes . . . user-driven imitation and remix have become highly valued pillars of contemporary participatory culture."[110] This focus on remixing and other forms of productive agency dramatically pushes back against the notion of viral transmission as an involuntary process. However, it is not the only way to make such a case. Even when people forward media content *as is*, they are still engaging in processes of selection and dissemination that are strongly agentic in character. To borrow a term popular in business circles, we can identify a *curatorial* agency that lies at the heart of selective forwarding practices.[111] In the curatorial metaphor, authorial voice comes not from creating symbolic content, but from assembling it and (re)presenting it to an audience. By choosing to pass on certain media artifacts over others in a crowded information environment, agents of selective forwarding actively help push the flow of media messages in the direction of their own interests, echoing Jenkins's adage that "materials travel through the web because they are meaningful to the people who spread them."[112]

Stressing the importance of curatorial agency within participatory media practice does not detract from or downsize the significance of creative or productive agency—without the latter, the former could not exist. However, a focus on selective forwarding, and the curatorial agency that it engenders, affords an opportunity to examine a much broader group of participants in the field of mediated political persuasion beyond the relatively smaller circle of professionals and amateurs who create their own symbolic content. In other words, we must consider not only the Krystal Lakes of the world who fashion political slogans like "America Was Never Great" (or, for that matter, the Donald Trumps of the world who craft media persuasion campaigns to advance elite power), but also all those who amplify and boost these messages by passing them along in both online and offline venues. Indeed, much of the activity discussed in this book involves the selective forwarding of persuasive political messages *as is*, such as changing one's Facebook profile picture to a social movement symbol, strategically tweeting links to news articles that raise awareness about select issues, and displaying mass-produced T-shirts, buttons, and bumper stickers that promote a favored electoral candidate or cause.

The notion of curatorial agency thus helps us to appreciate how joining broad-scale efforts to disseminate political messages can be conceptualized as potentially exercising a modicum of power and influence in a deeply mediatized political realm. Although many participants in such campaigns may not have a hand in producing or remixing the content itself, they do play a part in adding to its meaning and value by standing behind it and integrating it into the "face" of their public identities. Crucially, this added value is often seen as stemming from the authenticating power of ordinary people who choose to associate themselves with a message by presenting it via their personalized media outlets and channels. In other words, embodied personal endorsement is seen to lend an air of authentic grassroots support and enthusiasm to a political message, which may be instrumental in terms of its capacity to persuade. Thus, rather than viewing those who engage in selective forwarding as "dumb" connective links in a networked dissemination of ideas, we can recognize the complex relationships that form between individuals and the symbolic content they choose to incorporate into the public presentation of their social and political identities.

Furthermore, as much as the viral metaphor of involuntary contagion may be far from the reality, it is important to consider how it has impacted the way in which communicative power is conceived at a time when making content go viral has become a common frame. Although the notion of viral infection paints a picture of weak and vulnerable audiences that are susceptible to outside control, it also makes this control appear to be easily accessible to anyone who wishes to seize it for his or her own purposes. In other words, the viral media metaphor can be simultaneously empowering and disempowering for individuals, depending on whether one sees him- or herself as an active agent of dissemination or as a target. Moreover, this double-edge sword of the viral media metaphor seems to place an imposition on those who wish to see their political interests succeed in a mediatized political realm: *either infect the media or the media will infect you.* Presented with this choice, it is understandable that many citizens would try their hand at the media persuasion game in hopes of furthering their own political agendas. Thus, although we can uphold that the peer-to-peer spread of media messages is a process that is rich with human agency at every stage, we can also recognize how viral notions of mind contagion may act as a driving force in the popular uptake of the citizen marketer approach. Indeed, imagining those around you to be potential targets for your infectious message can provide an encouraging sense that you may in fact be able to make an impact on them and thus contribute to the broader shaping of public opinion. Although the viral contagion model is far too simplistic and mechanical to work as automatically as it suggests, it appears to serve as a useful fiction for many who wish to assert some small degree of control over the realm of political media and marketing that has historically poured down from above.

At the same time, it seems to demand a large leap of faith on the part of participants who must see themselves as meaningfully contributing to cumulative network dynamics that are, by definition, diffuse and aggregate in nature. The lack of immediate measurable feedback indicating the impact of one's own contributions to wide-scale projects of political promotion and influence presents a clear challenge for those who would take on a citizen marketing role, since it may be difficult to envision such connections between the individual and the mass. Yet as going viral becomes more of an accepted way of thinking about how microlevel acts of media sharing and spreading can add up to something much larger, the seductive appeal of citizen marketing will likely continue to gain force. Moreover, as discussed in the concluding chapter of this book, the recent growth of Big Data analysis promises to make peer-to-peer networked action increasingly quantifiable and legible, potentially adding further fuel to the citizen marketer approach. As the Facebook data analyst Eytan Bakshy notes in reference to the popular red equal sign profile picture campaign for marriage equality (discussed at length in Chapter 3), "for a long time, when people stood up for a cause and weren't all physically standing shoulder to shoulder, the size of their impact wasn't immediately apparent. But today, we can see the spread of an idea online in greater detail than ever before."[113] Notably, such an analysis says virtually nothing about the broader political consequences of a networked spread of ideas. Yet for those who adopt the approach of the citizen marketer, contributing to larger processes of message circulation is seen as a privileged pathway to empowerment in an age in which politics has become a fully mediatized phenomenon.

Outline of the Book

The citizen marketer approach to political participation may hold a great deal of promise for those who desire to remake the world in their interests, yet do these tactics make any difference? If so, how much? How could we even tell? From a practical standpoint, these may be the million-dollar questions, and *The Citizen Marketer* does not purport to have any easy answers. Rather, the primary purpose of this book is to qualitatively explore the experience of citizen marketing practice within lived social and cultural contexts. It is less concerned with gauging the precise extent to which citizen marketing translates into measurable political agency than with how this agency is conceived by social actors and what this can tell us about the shifting contours of political participation. By investigating the motivations and logics behind the act of pushing one's political viewpoints in public via media-based symbolic action, as well as how it has developed and intensified in response to key social, cultural, and technological changes, *The Citizen Marketer* charts the evolution of political participation in an age of mediatized politics, promotional culture, and viral circulation.

Furthermore, the book critically examines the tensions and uncertainties that arise from the imposition that citizen marketing and its lofty promises of empowerment place on those who wish to influence the political world. If we feel that we must outwardly promote our political beliefs and identities through continual participation in public mediated discourse, how might this potentially threaten the quality of our social interactions and contribute to a fractured and contentious public? In what ways might the imposition to take sides and show support for one competing set of viewpoints over another exacerbate ongoing trends of political polarization and partisanship? Furthermore, how might a shifting emphasis toward symbolic political goals risk a disconnection with the structural realities of the social and political world, and how might new forms of media and marketing literacy help avoid these pitfalls and potentially maintain such connections? *The Citizen Marketer* probes these questions and many others by putting the experiences of its practitioners under the microscope, teasing out the anxieties as well as the hopes of those who have leaped into the fray of participatory political promotion.

Interviews gathered from more than four years of qualitative empirical research form the core of much of this book, although Chapter 2 investigates the more historical context of citizen marketing, tracing a lineage that extends to the beginnings of political iconography and propaganda. However, the shift from authoritarian to democratic systems of government, along with the rise of the bourgeois public sphere, marks a key turning point in this history. As the badges and heraldry symbols of monarchic and despotic allegiance give way to the promotional spectacle of Western democratic elections, symbolic artifacts of political sentiment such as banners and buttons begin to offer new entry points for citizen participation. This emergent repertoire of collective action also becomes a significant aspect of organized social movements and protest groups, which similarly incorporated symbolic public displays into their efforts by the middle of the 19th century. However, although part of the story of citizen marketing emerges from the tradition of formalized collective assembly and public political spectacle, Chapter 2 also explores the influence of a more vernacular tradition of political message sending associated with cultural forms of expression such as dress. This culturally situated engagement with politics takes a revolutionary turn in the countercultural movement of the 1960s onward, as expressive style comes to serve as a symbolic public articulation of new political viewpoints and identities. Of particular significance in this history is the growth of the slogan-printed

T-shirt, a keystone object of citizen marketing that signals the arrival of mediated political expression into popular culture and the casual spaces of everyday life. After tracking the development of the "body screen" enabled by mediatized apparel and other conspicuous commodities, Chapter 2 turns to the digital body of Internet profiles—a set of platforms that has greatly multiplied the ways in which citizens can share political opinion with others. However, rather than emphasizing a transition from offline to online spaces of citizen marketing, Chapter 2 concludes by exploring the interconnections between corporeally embodied displays and the digital networks that often amplify and extend them in new and significant ways.

This historical context sets the stage for the following chapters, which draw on firsthand interview data to explore how citizen marketing is currently playing out in several key political contexts. Chapter 3 focuses specifically on the terrain of identity-focused social movements, using the extended case study of gay and lesbian activism to examine how citizen marketing practices are mobilized in strategic projects of mediated public visibility. Taking as its theoretical point of departure Hannah Arendt's notion of the space of appearance in the public realm, the chapter charts how citizens use media-enabled practices of self-labeling to announce the presence of their identities and attempt to influence perceptions of social and political reality. Through tactics such as posting identificatory profile pictures on their Facebook accounts, gay and lesbian citizens and their allies model a style of political advocacy that puts visible identities to work as visual rhetoric. Although this "coming-out" model may have particular resonance for the gay and lesbian community that has long sought to end its historical invisibility, it is now been adopted by a wide range of political constituencies that seek to challenge notions of who "the people" out there truly are. More critically, Chapter 3 also considers how public visibility campaigns may contribute to a flattening of differences, as social identities become branded with homogenized sets of symbolic artifacts. On the one hand, the packaging and branding of identity, long a sticking point in lesbian, gay, bisexual, and transgender politics because of its association with mainstream assimilation and the marginalization of intragroup difference, alerts us to the possible limits of visibility as a strategy for inducing social and political change. On the other hand, the grassroots nature of participatory coming-out campaigns also introduces opportunities for those who are marginalized in such efforts to become visible in turn, as strategies of parody and reflexive commentary are used to complicate unified and simplified displays of collective political identity.

Chapter 4 turns from the identity politics of social movements to the formal political realm of election campaigns. Drawing on the stories of citizens who voluntarily participated in the peer-to-peer marketing of candidates, as well as developments from recent election cycles in the United States and the United

Kingdom, the discussion explores the complex intersection between traditional top-down electioneering and grassroots political promotion that emerges from popular culture. In the process, the chapter examines how a fanlike cultural engagement with modern political brands fosters participatory forms of candidate promotion that extend far beyond a campaign's formal digital media outreach. Chapter 4 highlights how citizen marketers take on the role of cheerleaders for their political "teams," seeking to model enthusiasm and rally like-minded peers in hopes of achieving victory on the electoral playing field. This dynamic, the chapter argues, is becoming particularly important for outsider and insurgent candidates who depend on groundswells of grassroots momentum to advance their electoral fortunes. In addition, the analysis considers how some citizens circulate election-themed media to their peers as a means of sparking political discussions in everyday cultural spaces, which has the potential to contribute to the ideals of deliberative democracy and stoke the fires of partisan discord. Although citizen marketing practices may be helping to unify and energize groups of voters under the banners of contemporary political brands, they also risk furthering the dynamics of political polarization that threaten to divide the polity into self-enclosed and opposing camps.

This dynamic of partisanship also frames Chapter 5, which looks at how the citizen-level circulation of political news and information—particularly via digital social media—takes on an agenda-driven and marketing-like character. Drawing from the firsthand experiences of citizens who use platforms such as Twitter to link to news articles highlighting favored political issues, the chapter argues that the selective forwarding of journalistic content positions citizens in a public relations–like capacity, helping to draw favorable attention to some truths and narratives over others and thus fashion a strategic lens on reality. In the contemporary environment of information surplus, in which a great multitude of journalistic voices compete for the attention of a distracted public, the grassroots curation of news on social media along ideological lines serves as an entry point for citizens to participate in agenda-setting processes that are subtly, yet undeniably, persuasive. Furthermore, the increasingly partisan character of political information itself, from ideology-charged news and satire to activist-oriented citizen journalism, adds further fuel to the marketing-like orientation of citizens who strategically publicize and promote this content to their peers. Chapter 5 further explores how citizens navigate the tension between frameworks of political persuasion and civic education, with the latter being preferable to some who are reluctant to take on an explicit marketing-like role that they view as overly manipulative. The chapter concludes with an analysis of for-profit news sites that depend on the peer-to-peer circulation of articles for their financial livelihood, and addresses the broader consequences of journalistic content being shaped to go viral across networks of like-minded peers. Although the resulting focus on the personalized drama of Internet "cause célèbres" may open important spaces for citizens to encounter the political in culturally accessible and emotionally powerful ways, it also comes with attending risks of trivialization that recall the well-established critiques of political marketing.

The concluding chapter of *The Citizen Marketer* further interrogates the pressing debates and controversies that arise over the course of the book. Taking as its central theme the question of how we might develop a critical literacy of citizen marketing to assess its broader social and political consequences, Chapter 6 begins by returning to the slacktivism controversy that continually dogs this set of practices. Although there is little available evidence to support the theory that participating in symbolic, media-based actions becomes a one-to-one substitute for other forms of political participation, the converse theory positing that these practices will move citizens "up the ladder" of participation also demands critical scrutiny. As discussed in Chapter 6, a third possibility remains—that is, that citizen marketing practices may have little or no causal relationship with other forms of political participation whatsoever, and both organizationally active and organizationally inactive citizens remain at their respective levels of commitment even when engaging in media-based persuasion.

Yet beyond the issue of how citizen marketing may or may not impact other forms of political activity, what does the increasing focus on marketing-like practices mean for the future of democratic citizenship? It may potentially expand the circle of political participation by making it easier, more accessible, more casual, and even more fun—but at what cost? Indeed, the citizen marketer approach presents many notable risks: in addition to exacerbating political polarization, partisanship, and an ideologically driven relativization of knowledge, perhaps the most serious threat of citizen marketing's ascendance is that it might attenuate the connection between symbolic victories in the media and complex political realities on the ground. The challenge, then, for those who adopt these practices is to work to retain and strengthen connections between the symbolic and the material, and this requires a deep reflexivity about the nature of the political content we choose to spread to others and what we hope to achieve by doing so. By drawing critical attention to these practices and exploring the lines of tension that run through them, *The Citizen Marketer* compels us to heighten our awareness of our participation in the promotional culture of the contemporary political sphere, and to think more carefully through the meanings of what we push out into to the world via the plethora of media platforms that we can now personally curate and control.

In addition, Chapter 6 addresses how a critical literacy of citizen marketing practices must also include an enhanced awareness and consideration of the broader power structures that bear upon them, ranging from elite attempts to shape peer-to-peer political message flows to serve institutional agendas to gaps in technological access and skills that may reproduce inequality in citizen marketing's digital scenes. Furthermore, the discussion of critical literacy returns us to the profound tension regarding the relationship between citizen marketing and the neoliberal market ideologies that seemingly underwrite it. Here, I suggest a distinction that potentially can be drawn between an uncritical acceptance of marketing logics that replicates its hierarchical and managerial power structures, and a more critical acceptance that attempts to refashion the marketplace of ideas into a more democratized field of agonistic symbolic contestation. Importantly, this potential democratization of political marketing differs from the claim, often made by its professional practitioners and defenders, that political marketing is innately democratic because it caters to the public's wants and preferences through processes of opinion research and adjustment.[114] Although such a notion leaves in place the hierarchical power relations of elite managerial practice, a more radical reworking of the idea of political marketing calls on citizens to wrest a degree of control over the persuasion process through participatory media actions and interventions.

The task of navigating the tensions outlined above may be particularly urgent at the current juncture, yet such concerns are far from new. In Chapter 2, we will continue our exploration of the citizen marketing approach by charting its historical lineage over the decades and centuries, ranging from the formal public spectacle of mass democratic assemblies to the informal expressive practices of postwar popular culture. As we will see, public participation in the circulation of political symbols has long been a fixture of democratic life as well as a source of controversy and concern.

2

The Historical Lineage of the Citizen Marketer

Citizen Marketing in Two Waves

The year is 1840, and in cities across the United States, a new form of political spectacle is on the march. Tens of thousands of men parade down the streets, bearing cloth banners and silk ribbon badges that promote the candidacy of their favored presidential hopeful, William Henry Harrison. The image of the log cabin, a symbol of Harrison's purported salt-of-the-earth origins, is emblazoned on innumerable objects carried in the processions. So too is Harrison's portrait, as well as the inescapable campaign slogan, "Tippecanoe and Tyler too." Some parade-goers even roll giant balls made of buckskin through the streets that broadcast block-letter slogans in support of Harrison and his Whig Party. As the historian Roger Fischer notes, these pioneering promotional objects of the 1840 Harrison campaign can be understood today as "the remote ancestors of our modern billboards, bumper stickers and lawn signs,"[1]—not to mention the viral videos, image macro memes, profile pictures, and other digital media content that citizens now use to conspicuously promote their political preferences to their peers.

Far from being mere dazzling curiosities, the participatory visual extravaganzas of 19th-century American electioneering played a key role in marketing political candidates and parties to the public. According to Fischer, this campaign tradition depended on "vast numbers of rank-and-file supporters with enough enthusiasm to covet objects promoting a candidate and enough flamboyance to promote that commitment proudly on their lapels and elsewhere."[2] In other words, efforts such as the 1840 Harrison campaign were powered by the labor of early citizen marketers, who took it on themselves to forward their preferred candidate's message to their fellow voters through the public display of promotional material. Michael Schudson, in his historical treatment of the evolution of American citizenship, describes this mid-19th-century period of mass parties as being defined by the comradeship and loyalty of citizen-supporters, who avidly participated in election campaigns out of a strong sense of party affiliation.[3]

However, the importance of these rallies and processions for electoral promotion diminished in the era of 20th-century mass media, as movie reels, radio, and eventually television became the privileged means of marketing candidates. According to Kathleen Hall Jamieson, the "surrogate message carriers"[4] of 19th-century election spectacle were eventually made obsolete when politicians began using the new media of mass communication to take their message directly to voters. From this narrative, it would seem that participatory political marketing is a thing of the past—an archaic practice harkening back to a time when the bodies of supporters provided one of the few available platforms for political constituencies to circulate promotional messages among the populace. However, citizen marketing did not become extinct in the wake of the mass media revolution. On the contrary, it appears to be experiencing a renaissance in recent times, in both online and offline spaces, and in social movement as well as electoral-related contexts.

To what can we attribute this resurgence? As a "return to the grassroots," the historical arc of the citizen marketer bears substantial resemblance to Henry Jenkins's model of the three stages of cultural history.[5] For Jenkins, the first stage, folk culture, describes a now-bygone era in which ordinary people were active participants in their own culture (e.g., communal storytelling and music making, etc.). In the second stage, the mass culture of the 20th century, this grassroots activity largely disappeared as cultural products came to be produced by mass media professionals, and everyday people took on the role of passive and receptive audiences. Grassroots cultural participation is revived in Jenkins's third and final stage, convergence culture, as accessible peer-to-peer media technologies come to provide opportunities for members of the public to produce and distribute, as well as consume, the elements of their culture.

A similar trajectory seems to be occurring in the sphere of political communication. The participatory media environment of the Internet has paved the way for new forms of grassroots activity around all sorts of political phenomena, from the selective forwarding of news article links to the circulation of humorous political memes and viral videos. In a sense, we are witnessing a return of Schudson's effusive citizen-supporters among the most activated subsets of the citizenry, although, as we will see in this chapter and in subsequent chapters, they appear to be motivated less by traditional institutional affiliations than by the culturally situated identifications that have come to define the contemporary political landscape. Indeed, the advent of peer-to-peer networked digital media is only the latest development in a much larger set of trends that has rejuvenated the practice of citizen marketing in late capitalist democracies along the lines of cultural identity expression. Prior to the proliferation of peer-to-peer digital media, citizen-level political promotion was already making a significant comeback as the result of broad shifts associated with the postwar counterculture and related social movements. For the hippies, punks, Black Panthers, and many other youth-oriented groups, personal style became synonymous with announcing one's politics to the outside world. Sometimes this message-sending activity was abstract in character, as in the convention-shattering subcultural fashions worn to obliquely challenge the social status quo. Yet in other cases, it was pointed and direct, as in the popular uptake of buttons bearing symbols such as the peace sign and slogans such as "Make Love Not War."

This broadly defined second wave of citizen marketing can be distinguished from its predecessor in the sense that it brought political message sending into casual spaces of everyday life. Whereas the 19th-century era of banner-waving rallies and ribbon-festooned parades situated citizen marketing within formally organized, ritual-like contexts of public assembly, the postwar era has been characterized by a far more informal blending of politics and popular culture. Contrary to reports of its demise, the formal mass assembly remains an important (if sometimes neglected) feature of the modern political landscape. However, citizen marketing is now just as likely to be witnessed while walking through a shopping mall, driving down the highway, or logging onto a popular social media site. As part of a broader shift toward what scholars have described as the culturalization of politics,[6] citizen marketing of the postwar era has effectively blurred the boundaries between political expression and cultural self-expression. As we will see in this chapter, these developments have produced new hopes about expanding civic discourse to broad swaths of the public that have become disengaged from the formal political sphere, as well as new anxieties about the meaning and purpose of this discourse as it becomes enfolded more and more within the seeming frivolity of popular culture.

The aim of this chapter is to put the approach of the citizen marketer into a historical context by exploring its development and growth over the centuries. I begin by charting the emergence of citizen marketing's first wave in the early days of modern Western democracies, when mass participation in organized rallies, processions, and demonstrations first popularized the practice of publically displaying symbols of political sentiment for the purpose of peer persuasion. As I argue, the shift from authoritarian to market-like electoral structures in capitalist democracies such as Great Britain (and eventually the United States) helps to account for how the performance of citizenship within these societies came to be defined as including the expression—and promotion—of individual political preferences. This tradition of participatory political theater continued to evolve as the cloth banners and badges of the 19th century gave way to more modern accoutrements, such as printed placards and pins, even as the first wave of citizen marketing was eventually overshadowed by the mass media tactics of profession-alized political communication.

The historical lineage then turns to the postwar youth counterculture, when a new wave of citizen marketing shifted the practice of broadcasting one's politics in public from the scene of ritualized pageantry to more everyday sites of per-sonal self-expression. This second wave thus includes a wide variety of phenom-ena at the intersection of politics and popular culture. However, the discussion focuses in particular on the development of slogan and logo-printed T-shirts, which enjoyed a steady growth in popularity in the second half of the 20th century. Perhaps like none other, the story of the political T-shirt encapsulates the shift from formal to informal modes of citizen marketing, since it allows its wearer to effortlessly display the symbolic political content of signs and banners in more casual cultural settings. The printed T-shirt can be thought of as trans-forming the human body into an always-on media exhibition screen (limited to a two-dimensional, graphic format), affording opportunities for everyday peo-ple to disseminate favored political messages across the public visual landscape in new and novel ways. This "body screen" is paralleled in several simultaneous developments, such as the bumper sticker, which similarly transforms an auto-mobile into a media screen that can display slogans and other symbolic political content to onlookers as it moves through the spaces and places of everyday life. Although the printed T-shirt and its brethren enjoyed widespread popularity by the 1970s and 1980s, the mocking criticism leveled at this seemingly glib and trivial style of public political expression foreshadows the slacktivism debate that surrounds citizen marketing in its contemporary digital phase.

Indeed, the strategies of peer-to-peer political promotion that have arisen in the age of the Internet bear important resemblances to these material culture forbearers. The chapter's historical lineage concludes by exploring the growth of online participatory political campaigning, in which citizens utilize the "digital body" (to borrow a phrase from danah boyd[7]) of social media profiles and other accessible Internet platforms to circulate political content that they hope to see attain viral status. From the bulletin board and email campaigns of the 1980s and 1990s to the Facebook and Twitter "movements" of the early 21st century, the participatory culture of networked digital media has brought the practice of citizen marketing to an unprecedented level of public prominence. Although charges of superficiality and slacktivism have brought much controversy to this set of phenomena, it has also inspired a newfound optimism regarding how everyday citizens might take hold of the power of media influence traditionally held by elites and bring about desired political and social change. Rather than conclude the historical lineage of the citizen marketer by emphasizing a shift from offline to online spaces of communication, however, the chapter consid-ers how they continue to develop in concert as mobile technologies profoundly elide the boundaries between corporeally embodied and digitized spaces of political expression. Like the Harrison supporters of 1840, today's citizen marketers are turning to the spectacles of in-the-flesh embodiment and on-the-ground mass assembly as key means of publicizing their messages to others, only now the remediation of these spectacles via social media posts and YouTube videos extends them further into the recesses of popular culture.

The First Wave of Citizen Marketing: Mass Politics and Public Pageantry

The historical lineage of the citizen marketer begins in earnest with the rise of liberal democratic systems of government in modern Western capitalist societies, yet its ancestors are perhaps as old as civilization itself. From the beginnings of human history, political leaders, institutions, and ideas have been presented in symbolic form for the purpose of mass persuasion. Histories of political propaganda often begin in the age of antiquity,[8] when kings and emperors would glorify their rule, celebrate their military victories, and defame their enemies through carvings, paintings, and myriad other expressive forms. However, such displays of political authority tended to be strongly top down in nature. Indeed, the "classic" model of political propaganda describes how a society's elites secure and expand their power by symbolically communicating this power to their subjects, who are in turn positioned as passive spectators.

However, a far more participatory tradition of symbolic communication appears in the history of religious practice. Although religion is certainly not the same as politics, the two have been intimately intertwined for millennia, and religious symbols are frequently imbued with ideological meanings that relate to the social order of certain groups. In Emile Durkheim's landmark cross-cultural study of religious ritual, he draws attention to the use of visual emblems in building, strengthening, and reinforcing bonds within these social groups. For Durkheim, ritual emblems are defined as material objects that symbolize the group itself and thus provide a unifying focus of attention for its members:

> The emblem is not merely a convenient process for clarifying the sentiment society has of itself, it also serves to create this sentiment; it is one of its constituent elements. In fact, if left to themselves, individual consciousness are closed to each other; they can communicate only by means of signs which express their internal states. If the communication established between them is to become a real communion, that is to say, a fusion of all particular sentiments into one common sentiment, the signs expressing them must themselves be fused into a single and

unique resultant. It is the appearance of this that informs individuals they are in harmony and makes them conscious of their moral unity.[9]

As Durkheim suggests, bearing a ritual emblem not only reflects one's connection to the larger group, but also helps to actively forge it by allowing the individual to sense his or her unity with others in a tangible and visually perceptible form. In traditional Christian societies, for instance, wearing the emblem of the crucifix both expressed one's commitment to the prevailing social order (i.e., the authority of the church) and reinforced that very commitment, serving as constant reminder of the sentiment that it symbolizes to both its wearer and onlookers. In this sense, religious emblems can be understood as ideologically charged communicative objects that gain their collective persuasive force by being shared among individual members of a society.

This point has not been lost on the modern-day marketing industry, which has sought to exploit the connection between religious symbols and brand logos as emblems of group identification and unity. In the book *Why It Sells*, Marcel Danesi accounts for the success of corporate brand logos like that of Nike by explaining that "the objective of marketing today is to . . . get people to react to logos in ways that parallel how people once responded (and continue to respond) to sacred or mythical forms."[10] Although such comparisons between sacred symbols and modern brand logos may appear reductive, even gauche, it is important to bear in mind the powerful influence of symbolic religious practice on both commercial and political forms of media persuasion, particularly as they attempt to exploit the affective dynamics of collective identification for strategic purposes.

In addition to the iconography of religion, practices associated with the military also loom large in the historical lineage of the citizen marketer. Prior to democratic forms of rule, the practice of publicly displaying one's political allegiances in symbolic form was mostly limited to members of armies (in addition to members of the court), who would have a practical need to identify themselves on the field of battle through visual markers. For instance, during the Han dynasty in China in the second century AD, followers of a rebel group wore battlefield uniforms that included turbans made of yellow kerchiefs as a means of identification. This choice of color was not arbitrary; yellow symbolized the religious beliefs of the rebels and stood in contrast to the visual symbolism of the Han dynasty that they opposed.[11] A similar practice flowered in medieval Europe in the form of the heraldic tradition, which involved the display of elaborate visual emblems on shields, flags, and banners to mark military allegiances during combat.[12] One can identify persuasive elements of heraldic coats of arms: the illustrious designs glorified the powerful rulers and families that they represented while also distinguishing the identities of their armies.

However, although heraldic emblems can be seen as direct predecessors of modern political iconography, their intended purpose was not necessarily to express the inward political convictions of those who bore them. In monarchic, feudal, and other authoritarian systems of government, flying the colors and displaying the symbols of one's commanders was typically a matter of duty rather than choice. By contrast, democratic rule—at least in its ideal and uncorrupted form—is predicated on the act of individuals freely selecting their governmental representatives from a set of competing choices. Citizens in electoral democracies must therefore continually and actively *take sides* through their own will. The history of democratic governance is marked by persistent failures to live up to this ideal, and it would be foolish to pretend otherwise. However, this promise of choice—whether it is ever truly fulfilled—signals an important shift in how the role of ordinary people is conceptualized in democratic social orders. It is understandable, then, that the introduction of modern democratic power structures in the West coincides with the rise of popular symbolic markers of political allegiance. Once citizens had the (ostensible) ability to choose their leaders, they began to express these choices outwardly and symbolize their preference for one competing faction over another. The system of democratic elections does not in itself require its citizens to engage in the outward expression of political preferences, only the act of voting. However, from the beginnings of modern mass democracy, taking sides in the privacy of the ballot box has gone hand in hand with taking sides in the public spectacle of the streets.

To understand why this is the case, it is important to consider how the modern democratic process developed in the West alongside market capitalism, as well as the closely intertwined concept of the marketplace of ideas. Historians typically credit the latter concept to the famed English poet and political commentator John Milton, who in 1644 wrote the following as part of his work *Areopagitica*: "And though all the winds of doctrine were let loose to play upon the earth, so Truth be in the field, we do injuriously by licensing and prohibiting to misdoubt her strength. Let her and Falsehood grapple; who ever knew Truth put to the worse in a free and open encounter?"[13] Here, Milton lays out a rationale for the protection of freedom of expression, arguing that all ideas should be permitted to compete in public discourse because "Truth" will eventually win out over "falsehood." Although he does not employ the specific metaphor of the marketplace, Milton's vision of a competition of ideas in a "free and open encounter" suggests the influence of the broader social and economic conditions of the 17th-century Britain in which he lived, where early merchant capitalists competed for dominance in an open marketplace of goods. Importantly, such an ethos of free and open competition appears to have shaped the political as well as the economic imagination of Western capitalist democracies such as Britain, providing the conceptual foundation for both the modern electoral system and the system of contesting persuasive communication that came to surround it.

Indeed, to emerge victorious in democratic elections, the competitive parties must convince a majority of voters to choose their ideas over those of their opponents, using myriad forms of persuasion to do so. Without being able to freely "sell" their vision of rule to the populace, electoral candidates would be unable to secure the public support necessary to achieve victory and ascend to power. Margaret Scammell makes this point repeatedly in her work on political marketing in democratic societies, emphasizing that "marketing logic is inherent in the competitive structures of democracy; candidates/parties compete to influence publics with rival offers that are essentially promises, and as such are hugely reliant for credibility on reputation (image)."[14]

Furthermore, the freedom of expression that enables competing political factions to effectively market their images to the citizenry also allows members of the citizenry itself, including supporters of these various factions, to enter the field of political communication and contribute their own expression and commentary. A major source of this commentary is the press, an institution that has come to been viewed as essential for the functioning of democracy because of its role in circulating opinion as well as factual information about civic issues to a literate voting public.[15] In the early days of modern Western democracies, however, access to printing technology was a privilege held only by bourgeois elites, who often sought to influence the outcome of elections by publishing newspapers, pamphlets, and other materials that reflected their political preferences and ideological positions.[16] As the prominence of the 17th- and 18th-century partisan press in countries such as Britain and the United States underscores, political power within electoral democracies became irrevocably tied to the power to persuade and to communicate ideas—in seeming contrast to the power of brute force associated with more authoritarian forms of rule.

It is therefore no wonder that citizens without access to the printing press would likewise seek to make their political opinions and preferences known to their fellow countrymen, albeit using less technologically sophisticated means. Indeed, the earliest examples of citizen marketing activity that would be recognizable to contemporary observers date back to the origins of the modern democratic electoral contest. In Britain, in many senses the first Western democratic nation-state, freedom of political expression and the freedom to elect government representatives without interference went hand in hand, both being codified in the 1689 Bill of Rights that followed the Glorious Revolution.[17] In the ensuing electoral battles between the Whigs and the Tories for parliamentary dominance, supporters of both parties would take to the streets, as well as to the printing press, to publicly express their allegiances and help build support and enthusiasm for their favored side.

By 1755, symbolic political expression at the citizen level had grown into elaborate public spectacle and pageantry in Britain, as illustrated in a celebrated series of paintings by the artist William Hogarth.[18] Entitled *The Humours of an Election*, the paintings depict how voters would display their support for the Whigs and Tories by wearing orange and blue ribbons, respectively. For instance, in the painting *Chairing the Member*, a procession carries a victorious Tory parliamentary representative down the main street of town, while angry Whig supporters sporting blue ribbons attempt to interrupt the proceedings by waging battle with their orange ribbon–adorned rivals. In *An Election Entertainment*, Hogarth shows Tory representatives handing out orange ribbons to supporters at a large communal feast, offering a glimpse of how political institutions began to encourage public displays of allegiance through the distribution of symbolic material objects. Although Hogarth's satirical images undoubtedly take their fair share of creative license, these visual documents of 18th-century British elections help us to appreciate how early modern democracies fostered a culture of citizen-level political expression through spectacular forms of visual identification. At the same time as British newspapers published impassioned arguments about the competing "Truth" of the Whig and Tory visions of good governance, the political discourse spilled over onto the streets in a symbolic war of blue and orange, as citizens flaunted their support for their favored party by spreading their representative colors across the public visual landscape.

The electoral spectacle of Britain, however, proved to be no match for that of the United States, the nation that brought such a tradition to a new level of extravagance by the mid-19th century. Prior to the above-noted 1840 campaign of William Henry Harrison, the first rumblings of citizen marketing in the United States appeared in the context of grassroots public support for the Democratic presidential candidate and military hero Andrew Jackson. Nicknamed "Old Hickory" for his supposed strength in character and leadership, Jackson came to be symbolized by hickory poles and sprigs that supporters would bear in processions and rallies surrounding the 1828 and 1832 presidential elections. Around the same time, an entire industry of mass-produced trinkets sprang up around Jackson and other political figures that were similarly revered by the populace. For instance, Jackson's name and likeness were reproduced on domestic objects such as snuff boxes, thread boxes, flasks, tortoise-shell combs, plates, and lusterware pitchers, as well as displayable items like silk bandanas and cufflink buttons.[19] Significantly, these politically themed trinkets reflected the broader context of the growing consumer capitalism of the United States; as O'Shaughnessy argues, the tradition of American electoral spectacle is the result of "an entrenched belief in the commercial ethos."[20] By 1840, the log cabin–decorated material culture of the Harrison campaign institutionalized a tradition of participatory political marketing that would become the primary means of promoting U.S. presidential candidates throughout the 19th century. As noted at the start of this chapter, these early displayable objects of political identification and promotion, such as bandanas, badges, and banners, were wielded "as partisan weapons in political campaigns"[21] in formally organized events like street parades and rallies in public squares (meanwhile, more domestic trinkets were largely confined to the home as objects of private consumption and personal memorabilia).

This tradition of pageantry surrounding democratic elections would go on to influence the symbolic repertoire of social movement activism beyond the electoral context. For instance, in Britain, the routines of collective action underwent a dramatic shift from the 18th to the 19th century, as various organizations and movements began to mimic the orderly processions and symbolic adornments of parliamentary election rituals. As the historian Charles Tilly explains, public protest in Britain prior to the 1750s was far more unruly in character, typically involving direct actions such as the destruction of property, the invasion of land, the seizure of grain, and the disruption of ceremonies and festivals.[22] However, by the early 19th century, a new repertoire of collective action had solidified that included the first documented protest demonstrations and rallies. Like the 18th-century electoral pageants that proceeded them and served as inspiration,[23] these events featured a range of citizen-level symbolic expression through the use of spectacular material culture.

For instance, in 1834, a trade union in London organized a public demonstration in which each worker was instructed to wear "a crimson riband, one inch wide, between the first and second button-hole, on the left side of his coat"[24] to symbolize his membership and solidarity. In addition, the marchers carried banners to identify their particular trades, thus sending a clear and conspicuous message of the extent of support for the union's cause. Another procession in the town of Lancashire corresponding with a cotton textile strike in 1818 made use of slogan-covered flags, as well as the spectacular site of "coffins containing simulated weavers' skeletons."[25] As Tilly notes, "despite the implicit or explicit threat of disruptive action that accompanied many demonstrations, marchers often wore their best clothes, sported uniform ribbons, proceeded in ranks under the banners of their trades, and displayed collective self-control."[26] In other words, collective action by the early 19th century had become largely symbolic in nature, designed to communicate social and political viewpoints to powerful stakeholders as well as the broader viewing public. Similar sites were on display in the United States by the mid-1800s, closely following the British model. For instance, Susan Davis describes how striking laborers in 19th-century Philadelphia would parade through the streets bearing "tokens of identity, such as badges, sashes, ribbons, and banners"[27] that were used to "create and unify opposition to the masters"[28] as well as impress the broader community with an orderly public image that aided their cause.

Thus, the 19th century saw the flourishing of a ritualized, formalized, and organized style of participatory political communication that can be thought of as the first wave of widespread citizen marketing activity. Yet in historians' accounts of these various electoral and social movement–oriented traditions of public spectacle, the narrative invariably turns to one of decline by the 20th century. Davis, for example, describes how the tradition of laborers' parades largely disappeared in the 1900s as urban public space came to be redefined for commercial rather than civic use.[29] Furthermore, as noted previously, the focus of electoral political marketing in countries like the United States began to shift toward mass media outlets and away from participatory public spectacles in the streets. Jamieson marks the 1896 presidential campaign of William Jennings Bryant as a major turning point in this transition, noting that the candidate's frequent public speeches across the country, which were picked up by newspaper reporters and distributed to readers, essentially removed the need for "surrogate" citizen-level promotion via parade banners, kerchiefs, lapels, and bandannas.[30] This trend was greatly exacerbated by the development of broadcast media over the course of the 20th century, which enabled candidates and parties to transmit their persuasive messages directly to the voting masses in appealing audio and visual formats.

This is not to say, however, that the tradition of participatory political spectacle that peaked in the 19th century ever really died out. Rather, mass media coverage offered new ways of extending their public reach, although they were now only one element of a much broader mix of mediated public communication (and were thus less central overall). According to John Thompson, "as the new media of communication became more pervasive, the new forms of publicness began to supplement, and gradually extend, transform and displace" the publicness of co-present, face-to-face spectacle.[31] For instance, the social movement demonstration, whose stylistic form coalesced in the 19th century, continued to abound in the 20th century as a media-centric event. In these demonstrations, marchers made strategic use of large-type printed placards that were easily legible in photographs, film reels, and television (these mass-produced objects were a considerable improvement over their cloth banner predecessors, which were more expensive and time-consuming to make). Mediated images of civil rights marches in the United States are filled with these handheld placards, culminating in a veritable sea of printed signs at the famous 1963 March on Washington.[32] However, to follow Thompson's point, media coverage of this event was far more focused on broadcasting the speeches at the central podium than on the placard messages carried by members of the audience.

In the field of U.S. electoral politics, traditions of organized citizen-level public spectacle were in serious decline by the middle of the 20th century, although election-themed material culture continued to develop in new and innovative ways. As Fischer argues, elections of this era "inspired objects unsurpassed in creativity and variety by those of any other period, while at the same time campaign items . . . diminished steadily in importance as functional facets of mainstream elective politics."[33] Again, mass media outlets such as television emerged as the primary means to market candidates, parties, and ideas to voters, whereas promotional objects like cloth pins and buttons distributed and sold to supporters came to be seen as mere novelty curios. However, reflecting the thriving consumer culture of the period, the postwar U.S. electoral merchandising machine pumped out all sorts of such items, including a few that would ultimately go on to become widely popular platforms for all kinds of citizen-level political expression, above and beyond the electoral context: namely, the bumper sticker and the printed T-shirt. The first examples of the message-sending bumper sticker appeared in the context of the 1956 presidential election, when "more than fifty different varieties turned countless bumpers into miniature billboards promoting Eisenhower or Stevenson."[34] As for the slogan T-shirt, the earliest known examples date back to the 1948 campaign of Thomas Dewey and were made only in children's sizes[35]—a reflection of the T-shirt's status at the time as inappropriate outerwear for adults.

The story of the slogan T-shirt, which is central to the historical lineage of the citizen marketer, will be taken up in the following section. At this point, however, it is important to reiterate that despite such innovations in the electoral material culture of the mid-20th century, the participatory public pageantry that had surrounded their display for more than a century prior had been relegated to the periphery of a mass media–focused system of political marketing. To understand how citizen marketing practice evolved in this period, it is necessary to turn to other threads of its historical lineage—in particular, the history of vernacular political dress—which illuminate how citizen-level political expression would ultimately become intimately tied up in everyday practices of cultural self-expression.

The Second Wave of Citizen Marketing: Political Expression Goes Pop

In the previous section, we saw how symbolic political expression at the citizen level developed in Western democratic states in the context of routinized and formalized collective assemblies: electoral parades and processions, social movement demonstrations and rallies, and the like. However, in the 20th century, a new wave of participatory practices involving the public expression—and promotion—of political opinion began to appear in more informal venues of popular culture, broadening this activity from the context of the mass assembly to the arena of everyday life. This second wave was on full display by the 1960s, when youthful members of the hippie counterculture crystalized a style of hybridizing political and cultural expression that has continued to be widely influential. However, as the scholarship on political dress reveals, there are other noteworthy precedents for this landmark shift in the historical lineage of the citizen marketer.

One important example that can be considered transitional in this narrative involves the dress practices of the women's suffrage movement in Britain at the dawn of the 20th century. The Women's Social and Political Union (WSPU), a social movement organization whose primary cause was the fight for women's right to vote, adopted the colors of white, purple, and green to symbolize the organization as well as the broader movement that it helped to lead. In many ways, the expressive practices of the WSPU form a classic case study of citizen marketing as organized collective pageantry. For instance, in a 1908 demonstration in London, approximately 30,000 marchers proceeded through the streets wearing clothing, scarves, ribbons, hatpins, and brooches decorated in the WSPU colors, creating a mass visual spectacle that promoted the cause of women's suffrage to scores of onlookers. As Wendy Parkins explains, "through the use of fashion and specific colors, the suffragettes forged a public identity for themselves and introduced themselves and their cause into the sphere of political communication."[36] In other words, the public display of these colors became a coded, yet highly legible, political statement about women's suffrage that was well understood within British society at the time.

However, although the WSPU primarily encouraged its members to wear the colors at official public occasions to "help to make it understood that women are standing together and supporting one another in the great fight in which we are engaged,"[37] this expressive activity spilled over into more informal cultural spaces. As Lisa Tickner documents, the symbolic colors of white, purple, and green could still be witnessed in the streets of London after the Hyde Park demonstration had concluded, with one observer commenting that this sight served as a constant reminder of the suffragettes' cause.[38] Indeed, the WSPU had launched a brief fashion craze among London women at the time, as the clothing and accessories procured for demonstrations entered the wardrobes of the suffragettes and became part of their self-expressive practices more broadly.

This use of personal style to make statements of inward political convictions became more prominent over the course of the 20th century, corresponding with a broader shift in the social conventions of Western dress. According to the fashion sociologist Diane Crane, dress practices in the 19th century were largely focused on the indication of social class, whereas in the 20th century, dress became more associated with individual lifestyle choices and the construction of self-identity. As Crane posits, this sartorial shift reflects the rise of

postindustrial consumer culture in the West, which gradually replaced traditional social categories with fragmented lifestyles organized around categories
of leisure and consumption.[39] Theorists of modernity such as Anthony Giddens
similarly characterize the "self" in the 20th century as a continual project of individual construction—a project aided by the emergence of a lifestyle-oriented
consumer culture that came to offer a seemingly endless array of tools for conspicuous self-expression.[40] Although many of these expressive lifestyles did not
necessarily carry a recognizable or legible political inflection, others did, particularly those associated with the subcultures of marginalized identity groups.

For instance, Elizabeth Wilson describes how "oppositional fashions" that
deliberately diverge from social convention have been used "to express the dissent or distinctive ideas of a group, or views hostile to the conformist majority."[41] As an early example, she cites the Parisian Bohemian subculture of the
1830s, whose members used a distinctively anachronistic form of dress to
express their rebellious attitude toward the dominant ideologies of French society. This pattern of oppositional style became more common in the mid to late
20th century, however, when numerous subcultural groups emerged as a result
of social fragmentation. One particularly significant example noted by Wilson is
the American Black Power movement of the 1960s and 1970s, which made use
of Afrocentric clothing and hairstyles to send a "conscious and deliberate message"[42] of black pride that directly confronted the racism of the white majority.
Importantly, this style of dress was not limited to the marches and rallies orchestrated by social movement organizations like the Black Panthers, but was also
incorporated into the self-expressive practices of everyday life. By simply walking down the street with "Afro" haircuts and dashikis, young blacks of this period
broadcasted lucid political ideas to their peers that openly challenged bigotry
and racial oppression.

For Wilson, the strongly political character of Black Power subcultural dress
was an exception rather than the norm, since the symbolism of oppositional fashion has tended to be more implicit than explicit. In the 1960s hippie subculture,
for example, wearing tie-dye shirts and bell-bottom jeans symbolized one's affiliation with this youth-oriented identity group as well as a broad rejection of the
social status quo, but these articles of clothing alone hardly articulated a clear set
of political viewpoints (Dick Hebdige, the preeminent scholar of youth subcultural style, documents how the punks of 1970s Britain similarly used this sort of
oppositional dress as coded "symbolic refusal" of mainstream social conventions
and ideologies[43]). However, as members of the hippie subculture began to incorporate elements drawn from the tradition of formalized political promotion and
pageantry into their personal repertoires of rebellious self-expression, they modeled a far more legible and conspicuous form of political statement making in the
leisure spaces of popular culture. In particular, the slogan-printed button, which

had for decades been a fixture at electoral campaign events and social movement rallies and marches, was adopted by many hippies as an everyday accoutrement. For instance, one scholarly account of the typical hippie appearance from this period notes "peace buttons such as 'Make Love Not War'" alongside items such as paisley shirts, beads, sandals, and flowers.[44]

This sort of intermixture of political and cultural expression was a hallmark of the 1960s countercultural movement more broadly, which centered on rock and folk music that often featured antiwar lyrics and other explicitly political content. Indeed, the practice of injecting political messages into a broader repertoire of cultural self-expression was epitomized by many of the hippie movement's most renowned cultural figureheads. For example, John Lennon, the world-famous Beatle and hippie par excellence, deliberately sought to promote a peace message in all aspects of his life, including numerous media appearances and creative works during the late 1960s and early 1970s. As he explained to an interviewer in 1969, "Henry Ford knew how to sell cars by advertising. I'm selling peace at whatever the cost. Yoko and I are just one big advertising campaign."[45] This adoption of marketing discourse by Lennon was not just metaphorical: for instance, a photograph taken in 1971 shows Lennon lying in the grass while wearing a T-shirt printed with a large peace sign and the slogan "Come Together," taken from one of his popular antiwar songs with the Beatles.[46] Here, Lennon literally transforms his body into a billboard for the antiwar movement, even while doing something as quotidian as casually reclining in a field. As images like this attest, Lennon was acting as a kind of prototype for a new form of culturally situated citizen marketer—one who incorporates persuasive political messages into a range of everyday self-expressive practices. It is not a stretch to surmise that Lennon's high-profile peace-promoting activities set an influential example that many future citizens would follow, both within the hippie movement and beyond.

Lennon's peace sign–adorned T-shirt is also one of the earliest documented examples of a politically oriented graphic T-shirt worn by adults. As noted previously, the first T-shirts printed with politically oriented content were made exclusively in children's sizes as promotional campaign items of the major U.S. electoral parties. In the 1940s and 1950s, printed T-shirts were a common item for children, but were not considered respectable adult dress. Even wearing a plain white cotton T-shirt as outerwear (as opposed to underwear, its original purpose) was considered a rebellious act of oppositional dress. Indeed, the style was famously adopted by teenage males within the motorcycle and drag-racing subcultures, and the exemplary white T-shirts worn by Marlon Brando in *The Wild One* from 1953 and James Dean in *Rebel without a Cause* from 1955 are often credited with establishing the T-shirt itself as a cultural symbol of youthful rebellion around the world.[47] However, it was the counterculture of the 1960s that first embraced the graphic printed T-shirt as a mediated public platform of self-expression, which essentially extended the protest button into a larger and more easily legible format. In addition to the ubiquitous peace sign, iconic symbols of leftist protest such as the portrait of Marxist revolutionary Che Guevara[48] and the logo of the Black Panther party[49] could be seen emblazoned on the chests of youthful activists by the end of the pivotal decade of the 1960s.

As noted at the beginning of this chapter, the rise of the printed political T-shirt marks a critical turning point in the historical lineage of the citizen marketer—a moment in which citizen-level symbolic political expression truly began to flourish in the realm of popular culture and outside the parameters of formalized political pageantry. The T-shirt's status as inexpensive leisurewear underlines this point. Although it can be worn at organized political assemblies in a similar way to the badges, ribbons, and sashes of earlier eras, the T-shirt is equally at home in the casual spaces of everyday life. The protest slogan buttons of 1960s hippie fashion may have initiated this trend, but the T-shirt has proven far more durable as a platform for political expression and promotion in the ensuing decades. Unlike the button, an item that is not typically worn in everyday dress, the T-shirt blends seamlessly into popular repertoires of personal self-expression via consumer culture.

By the 1970s and early 1980s, adult T-shirts imprinted with all kinds of images and slogans—promoting everything from rock bands to liquor brands—became a fixture of the popular culture landscape. A quote from the *Washington Post* from 1982 gives a sense of the developing cultural terrain: "Today, nearly every town, organization, sports team, company and school has its shirts. No longer do firms pay people to wear sandwich boards advertising their wares; instead, consumers spend their own money to buy shirts, and other items, that tout favorites: Budweiser, I heart New York, Pac-Man or Surf Naked."[50] Crane describes the printed T-shirt as a communication platform that "epitomizes postmodern media culture,"[51] serving as a personalized, embodied screen that can be used to broadcast—and promote—virtually any message imaginable to the outside world. The adoption of this culturally situated platform for political messaging signals the direction that citizen marketing would eventually take as the available formats and technologies of media participation began to multiply.

In the 1970s, some of the first identifiable political T-shirt trends began to appear in American popular culture, that is, politically themed designs that were distributed widely enough to become known for *being* T-shirts. Reflecting the by-then long-term association of T-shirts with youthful rebellion and protest, such designs were typically centered on left-wing, youth-oriented social movements. For example, T-shirts calling for the release of Angela Davis, the imprisoned activist involved with the Black Panther Party, were sold widely both in the United States and abroad from 1970 to 1972;[52] the design even featured a

photographic portrait of Davis, taking advantage of newly invented fabric trans-
fer technologies to add a level of urgent realism to the shirt's message. The feminist movement also launched several iconic T-shirt designs in this decade, such
as the humorous shirt reading "A Woman without a Man Is Like a Fish without
a Bicycle."[53]

Another noteworthy trend in the 1970s involved the appropriation of political T-shirts by the nascent gay and lesbian movement. During this period, the
severely marginalized position of gay and lesbian activists within the broader
society inspired them to develop innovative ways of communicating their message in public when official media channels were mostly closed off. Seeking to
make its voice heard, the gay and lesbian movement turned to "alternative channels . . . with low entry barriers."[54] This effort was exemplified by the wave of
T-shirts that appeared in 1977 targeting the anti-gay activist Anita Bryant, who
at the time was spearheading a Florida ballot initiative to fire gay and lesbian
teachers from public schools. For example, the popular "A Day without Human
Rights Is Like a Day without Sunshine" design parodied Bryant's famous tagline
as a television spokesperson for Florida orange juice to humorously promote the
gay rights cause.[55] As design trends such as the orange juice–themed anti-Bryant
shirts proliferated through wide-scale merchandising efforts, their potential to
impact the broader public visual landscape was magnified.

Perhaps no one better exploited this concept of T-shirts as a platform for
the promotion of political opinion than the British fashion designer Katharine
Hamnett. She was not the first fashion industry figure to produce T-shirts
with politically charged content—that distinction likely belongs to Vivienne
Westwood, who became known in the mid-1970s for selling punk shirts that
featured oblique protest designs juxtaposing profanity, pornographic imagery,
swastikas, and portraits of Queen Elizabeth II.[56] However, Hamnett's more
direct sloganeering revolutionized the use of T-shirts for broad-scale media-
centered activism. Already a well-known designer by 1983, Hamnett became
world famous that year for sending T-shirts down London runways printed
with large block-letter environmentalist slogans like "Stop Acid Rain" and "Stop
Killing Whales."[57] As she explained in a press interview, "fashion gets too much
publicity. Around 1983, I realized I could use the coverage we were getting to
promote green causes, and put over seminal ideas to people who might not
otherwise be receiving them."[58] In other words, Hamnett envisioned her mass-
produced T-shirts as a grassroots political marketing campaign, exploiting the
media coverage her fashion label received to inject environmentalist (and other
left-progressive) advocacy messages into the public discourse. In another interview, Hamnett commented that her T-shirt designs were "an exercise in manip-
ulating the media, in getting a message across. . . . Fashion gives you a voice.
I have the intelligence to use that voice."[59]

The use of this T-shirt "voice" was most dramatically modeled in a publicity stunt that Hamnett orchestrated in 1984 while wearing one of her own slogan shirts. While attending a reception for British Fashion Week at the prime minister's residence at 10 Downing Street, Hamnett greeted Margaret Thatcher by revealing her T-shirt, which read "58% Don't Want Pershing" in unmistakably legible oversized black letters.[60] The slogan referenced British public opinion about arms buildup (specifically the purchase of American Pershing missiles by the U.K. government), and Hamnett was ostensibly using her T-shirt to directly lobby Thatcher about the issue. However, as she later explained, "I just wanted a slogan that would show up well on 35mm film."[61] Indeed, photographs of this unique encounter between Hamnett and Thatcher were splashed across newspapers and the media more broadly, giving her T-shirt message widespread visibility. The fact that Hamnett used her T-shirt to inject an advocacy message into a social setting where such sloganeering is normally not expected to occur (i.e., a meet-and-greet, high-society soiree) underlines how this medium has been used to blur the boundaries between political and cultural expression. As Hamnett's slogan T-shirts became popular worldwide, this more casual form of citizen marketing activity proliferated more and more among the ranks of everyday citizens.

Along with its growing popularity, the political T-shirt notably became a frequent target of satire and criticism. For instance, in 1978 the author and humorist Fran Lebowitz commented that "clothes with pictures and/or writing on them . . . are an unpleasant indication of the general state of things. . . . If people don't want to listen to you, what makes you think they want to hear from your sweater?"[62] Here, Lebowitz jokingly articulates a concern about the intrusive—even confrontational—quality of casual sloganeering that would continue to surround culturally situated forms of citizen marketing in the coming years. Another common complaint is exemplified in a journalistic piece covering Hamnett, which mockingly suggests that "apparently, her customers and those who read about her in the papers are so pitifully ill-informed that they depend on Hamnett for the planetary news headlines."[63] The implication here is that T-shirt-based expression is a dumbed-down and ultimately vapid form of political discourse—a concern that has multiplied as citizen marketing activity has expanded to more sites of popular culture.

However, whereas some derided the political T-shirt phenomenon as a tedious frivolity, others feared its potential power. In a sign of the political T-shirt's growing international prominence, some authoritarian governments began to ban the items out of concern for their persuasive impact. For instance, in 1987, the apartheid-era South African government issued a ban on wearing T-shirts, badges, and stickers "depicting a slogan protesting or disapproving of the detention" of political prisoners, part of a broader push to stifle the anti-apartheid movement in the country.[64] In 1991, the ruling Chinese Communist Party banned the sale of widely popular so-called culture shirts, which featured vague but politically charged slogans such as "I'm Fed Up, Don't Bug Me" and "Sick and Tired" and were interpreted by the government as the work of "antisocial elements."[65] As examples such as these illustrate, the popular-culture form of the T-shirt was beginning to be taken seriously around the world as a powerful— even potentially dangerous—agent of social and political persuasion.

Indeed, whereas some governments acted to ban these items, many political institutions around the world embraced them as tools of electoral promotion. One of the earliest examples of election T-shirts for adults appeared in the 1972 Australian federal election, when the Labour Party led by Gough Whitlam inundated the country with shirts and other items printed with the slogan "It's Time," which cheekily referenced the long-running incumbency of the rival Liberals.[66] In Brazil, the hypercompetitive multiparty system that emerged in the 1980s after a long dictatorship gave birth to a vibrant culture of election-oriented political expression in the streets; along with items like stickers and lapel pins, T-shirts became a highly visible fixture of Brazilian presidential campaigns and were used prominently by the labor-oriented PT party to promote the Lula candidacy.[67]

In the United States, the election T-shirt had reached such a level of cultural ubiquity by the time of the 1992 contest between Bill Clinton and George H. W. Bush that the Fashion Institute of Technology Museum in New York City devoted an entire exhibition to it. Entitled "V-O-T-E: The 1992 Election in T-shirts,"[68] the exhibition featured official logo-printed T-shirts produced by the Clinton and Bush campaign organizations alongside satirical T-shirts sold on the commercial market. In an accompanying essay, the curators, Richard Martin and Harold Koda, eloquently explain how these T-shirts represent an important innovation in participatory democratic communication: "In 1992, Americans are demonstrating their social contract and democratic duty in being unafraid to express themselves on the issues of this election . . . the intensity of this year's T-shirts gives us reason to believe that voters are ready to speak out, ready to exchange ideas, ready to criticize."[69] Here, Martin and Koda's observations about the cultural landscape surrounding the 1992 U.S. election seem to capture the broader zeitgeist of citizen marketing's proliferation in popular culture by this period. As their exhibition intended to illustrate, grassroots political expression at the citizen level was making a significant comeback long after its 19th-century heyday, now encompassing both the official promotional outreach of political organizations and an unofficial, often satirical, material culture of vernacular political commentary.

The growth of the political T-shirt in the latter half of the 20th century from children's novelty to ubiquitous global force of grassroots media illustrates the ascent of citizen marketing in its culturally situated second wave. Similar trajectories can be traced in the popularization of other politically themed expressive objects of postwar consumer culture, such as the bumper sticker. One survey of American automobiles conducted in 1996 concluded that roughly 40 years after they were first introduced, "the incidence of bumper stickers as a form of political and social expression appears frequent and significant."[70] However, imprinted material objects such as T-shirts and bumper stickers have obvious limitations as platforms of public communication, since they can only be seen by those who happen to be in their immediate physical vicinity (unless, of course, they are remediated via photography and video). The relatively minimal reach of sartorial and automobile "broadcasts," in comparison to traditional mass media, likely accounts for why they have received little attention from scholars of political communication. Beginning in the 1980s, however, new participatory media technologies began to offer citizens the promise of a much wider audience for their political promotion and persuasion efforts. Furthermore, whereas acquiring and displaying material objects for political messaging purposes often requires notable time and effort (as well as financial resources), spreading political opinion online can be done at an infinitesimally minimal cost, which has undoubtedly played a key role in its widespread popularization. Thus, as the digital age dawned, citizen marketing would enter a period of intensification and multiplication that would compel a far more serious consideration of its power to impact the broader political landscape.

Citizen Marketing in the Digital Age: Political Expression Goes Viral

The notion of using peer-to-peer digital networks for grassroots political advocacy and campaigning is nearly as old as the Internet itself. Prior to the widespread popularization of the World Wide Web in the mid-1990s, early conferencing networks like USENET and the WELL, along with online bulletin boards and email lists, were hotbeds of social and political expression and had numerous ties to left-wing counterculture. For instance, the WELL, which stands for Whole Earth 'Lectronic Link, was cofounded in 1985 by Stewart Brand of the environmentalist Whole Earth Catalog and was conceived from the outset as an alternative communication space for self-styled radicals. In *Media Virus!*, Douglas Rushkoff posits that "there are many different sorts of opinions expressed in the conferences of the WELL," yet "the overall sentiment is a highly-progressive, pro-environment, pro-psychedelics, pro-tolerance one."[71]

Indeed, Rushkoff's early celebratory vision of networked digital media as having the ability to reorient the direction of society and politics—discussed in Chapter 1—is steeped in the countercultural orientation of late-1980s and early-1990s Internet culture. Before the commercial browser brought the World Wide Web to millions around the world and launched the "dot com" boom, the Internet was largely conceptualized as a noncommercial alternative to mainstream media that empowered users to share information and ideas that could not be found elsewhere. Reflecting this vision of the Internet as a radical experiment in media democratization, early online platforms like the WELL and USENET attracted communities of users who saw themselves as challenging the perceived hegemony of corporate and government-controlled mass media. Thus, Rushkoff writes that "facts deemed too controversial for the news find their way onto bulletin boards and USENET groups when anyone, anywhere, with a computer decided to post them."[72] As an example, he quotes John DiNardo, a frequent contributor to the USENET Activists group, who included the following plea to close his politically charged posts circa 1992: "The American Public is evidently in dire need of the truth, for when the plutocracy feeds us sweet lies in place of the bitter truth that would invoke remedial action by the People, then we are in peril of sinking inextricably into despotism. So, please post the episodes of this ongoing series to computer bulletin boards. . . . The need for concerned people alerting their neighbors to overshadowing dangers still exists, as it did in the era of Paul Revere."[73]

This use of participatory digital networks to selectively forward news stories to one's peers that may not be encountered otherwise—an activity discussed in detail in Chapter 5—signals an important way in which the Internet significantly expanded the scope of citizen marketing activity. In contrast to material objects of political expression that can only display a limited amount of static media material (slogans, logos, etc.), online posts allow media nonprofessionals to circulate far more complex types of media content to their peers, including in-depth journalistic articles. Whereas a slogan on a T-shirt or handheld placard might concisely spotlight a piece of information that one wants others to know about (e.g., Katharine Hamnett's "53% Don't Want Pershing"), digital technology is far more flexible and expansive in its communicative capabilities. Moreover, the fact that a left-progressive online venue like USENET was known for featuring posts that can be characterized as advocacy journalism (Rushkoff further cites stories about prison corruption and the U.S. military–industrial complex as examples[74]) underlines how this practice has tended to blur the goals of informing one's peers and persuading them.

A second major shift in citizen marketing as it entered the digital age is further indicated in DiNardo's call for his readers to repost his missives on bulletin boards across the Internet. Here, we can glimpse how the networked character of the Internet began to allow for a much faster (and potentially far more widespread) dissemination of politically oriented media material at the citizen level. Whereas the "broadcasts" of displayable material objects cannot extend beyond one's immediate physical surroundings without further remediation, forwarded content on peer-to-peer digital networks is nearly infinitely replicable and spreadable. The early Internet culture of bulletin boards, conferences, and email lists was well aware of these network dissemination effects. Indeed, some of the first self-consciously viral media of a political nature appeared during this era. In 1993, for instance, a series of messages labeled "Virus 23" circulated on a popular email list known as FutureCulture, each featuring self-reflexive variations on the theme of infecting the culture with radical ideas (or memes, in the Richard Dawkins–derived terminology preferred by many Internet users[75]). One example of a Virus 23 email claimed to infect its readers with notions of various countercultural religious movements and contains passages like "this text is a neurolinguistic trap, whose mechanism is triggered by you at the moment when you subvocalize the words 'Virus 23,' words that have now begun to infiltrate your mind in the same way that a computer virus might infect an artificially intelligent machine." Another Virus 23 email professed to spread the meme of the coming ascendency of black Americans in a deliberately radical fashion.[76] Although these early examples of intentional peer-to-peer persuasion through online channels may appear obscure or obtuse, they foreshadow the veritable cavalcade of social and political influence campaigns that would flourish in the social media age.

As the example of FutureCulture highlights, email provided Internet users with one of the first accessible platforms to selectively forward digital messages to peers for persuasive political purposes, long before the advent of commercial social networking sites such as Facebook and Twitter. One particularly well-documented example of a politically charged email meme involves the culture-jamming efforts of the activist and entrepreneur Jonah Peretti (who went on to cofound both *The Huffington Post* and *Buzzfeed*). In 2001, Peretti initiated an email exchange with representatives from the footwear company Nike about his wish to purchase sneakers customized with the word sweatshop. Although his request was rejected, the emails—which cleverly drew attention to global labor abuses and corporate malfeasance—were forwarded thousands of times and eventually received coverage from mass media outlets.[77] This coverage enhanced the exposure of Peretti's political critique even further, yet the peer-to-peer selective forwarding of the emails provided opportunities for citizens to directly participate in the spread of the Nike sweatshop meme. In recounting this incident as an example of what he calls "new media power,"[78] Bennett underscores how the introduction of the Internet has fueled new optimism about the ability of ordinary citizens to impact the sphere of political communication and counter the dominance of traditional elites.

With more and more political uses of networked digital media proliferating in the early 21st century, this optimism has continued to grow. For instance, the use of peer-to-peer networks to mobilize large-scale offline actions—first demonstrated in dramatic fashion in 2001 during the SMS text message–powered uprising to overthrow the Philippines president Joseph Estrada[79]—has led to bold claims about the Internet as the progenitor of new global revolutions. However, this sort of logistical planning activity, defined as "e-mobilization" by Jennifer Earl and Katrina Kimport,[80] has a somewhat different tenor than most of the peer-to-peer political persuasion activities considered in the chapter, as well as this book as a whole. Although digitally forwarding the times, locations, and instructions for offline actions does indeed function to promote or advertise these actions, the intended audience may be already-committed members of an organization who simply require the next bit of practical information to coordinate their efforts. The practice of citizen marketing, by contrast, largely focuses on selectively forwarding ideas and opinions to those who are not yet fully behind them to influence their thinking. However, information that appears to be purely logistical (such as the text message "Go 2 EDSA. Wear Black" in the case of the 2001 anti-Estrada protests[81]) may have the potential to encourage, as well as coordinate, participation in offline protest actions. In the Egyptian uprising of the 2011 Arab Spring, one of the most prominent examples yet of "e-mobilization," the "We Are All Khaled Said" Facebook page was described as serving "both as a kind of central command and a rallying point—getting people past 'the psychological barrier.' "[82] By spreading information that a massive protest of tens of thousands was immanent, Egyptian citizens helped convince their more reluctant peers that they would be protected by the sheer size of the crowd and that the movement to overthrow the president, Hosni Mubarak, was large enough to have a chance at success. As this case illustrates, the coordination of offline protests through digital means may contain elements of peer persuasion, even if it may not be the exclusive intent or purpose. Thus, without drawing any hard and fast lines, we can say that citizen marketing has certain overlaps with the growing phenomenon of "e-mobilization," yet is not necessarily an equivalent category of analysis.

Another type of Internet activity that overlaps to some degree with the peer persuasion of citizen marketing is online petitioning, which Earl and Kimport classify as an "e-tactic," in contrast to e-mobilization. As the authors describe, participatory online petition campaigns were first popularized by sites like PetitionOnline, which launched in 1999.[83] Many similar sites have appeared since, such as Change.org, which boasts more than 71 million participants on its website, as well as the globally minded Avaaz.org, whose site claims "34 million members in 194 countries." Even the U.S. government joined the online petition craze in 2009 with the site We the People, which was introduced with

the premise that Obama's White House would formally respond to any petition that reached a certain threshold of online signatures.[84] The structure of We
the People underlines how online petitions are a form of participatory political
communication that formally targets an audience of elite decision makers, as
opposed to an audience of fellow citizens. As Karpf emphasizes, these online
petitions are similar to their pen-and-paper forbearers, which have long been
used by advocacy organizations to pressure government officials to alter their
decision-making processes.[85] Although critics like Stuart Shulman have pointed
out how the flood of e-petitions may simply be ignored by overwhelmed politicians,[86] Karpf counters that they often help organizations get face-to-face meetings with these officials and add rhetorical strength to their behind-closed-doors
lobbying efforts.

In this sense, the tactic of the authority-targeting petitions diverges somewhat from the peer-to-peer influence model of citizen marketing. However, as
online petitions are promoted and circulated via email, Facebook, Twitter, and
other digital platforms (as sites like Change.org and Avaaz.org indeed facilitate), information about the issues that they seek to redress are also disseminated to an audience of peers. Although the stated purpose of an online petition
is to influence elite power brokers to take a specified action, even those which
fall short of this goal may have a secondary or peripheral effect of increasing
public awareness about select issues in ways that may hold the potential of paying future political dividends. Scott Wright highlights this point in a study of
e-petition creators in the United Kingdom, finding that they had many ways
of defining the "success" of their e-petitions that went beyond the traditional
measure of achieving stated objectives and included a nuanced consideration
of the broader public impact of their circulation.[87] Thus, like e-mobilization, the
widely popular e-tactic of online petitioning contains certain notable traces of
the peer-to-peer citizen marketing model, although this may not be its primary
focus or intention.

As Karpf notes, online petitions have faced particularly strong derision for
fostering a form of political participation that requires little time and energy
in comparison to traditional forms of organized activism: "The fear is that the
resulting waves of minimal-effort engagement hold long-term costs for the public sphere, either by further dispiriting the issue publics who find their online
petitions and e-comments ignored, or by crowding out more substantive participatory efforts."[88] The notion that online political expression threatens to take the
place of traditional political participation—what Henrik Christensen calls the
substitution thesis[89]—will be taken up later in Chapter 6. At this stage, however,
it is important to note that the current e-petition controversy illustrates how
the lowered participation costs of online digital media have heightened fears
about the enervation of grassroots political activism. Prior to the popularization

of online petition sites like Change.org and Avaaz.org, there was little public concern about the deleterious effects of the petition as a political tactic more broadly. Yet now that signing a petition has become as simple as clicking a few buttons on a laptop or smartphone screen, many cannot help but wonder how this might cheapen the meaning—and perhaps even the force—of political participation in the digital era.

Indeed, the slacktivism debate has continued to build as more forms of online grassroots activism have made their way into the political and cultural landscape. With the advent of social networking and microblogging platforms in the early 21st century, new formats of political expression such as the profile picture, the hashtag, the image macro meme, and the viral video have led to much hand-wringing over the questionable role of networked digital media in contemporary political participation. It seems that every time a peer-to-peer online advocacy effort achieves a notable level of virality, a new round of editorials and think pieces pops up that decries the end of "real" political activism in the social media age. This is typically followed by a rash of response pieces that defend online advocacy tactics as making a meaningful impact on political discourse in a democratized, participatory fashion. For instance, #BringBackOurGirls, the prominent 2014 hashtag campaign that sought to foment public outrage about the human rights abuses of the Boko Haram group in Nigeria (discussed in Chapter 1), was variously described by observers in the press as "pathetic . . . a substitute for making a real difference"[90] and a tool helping to "proliferate worldwide awareness . . . the hashtag may just have been the best thing that you could have done to help."[91] As of late, this cycle shows no signs of abating. Rather, it seems that each viral act of citizen-level political expression becomes grist for the mill for our collective hopes and fears about the radically transformative potential of Internet technologies.

Conclusion: Looking to the Future, Learning from the Past

Although the slacktivism debate continues to rage with the appearance of each new politically charged meme and hashtag, the terms of this debate can be reoriented by considering the broader historical lineage of citizen-level political promotion. Much of the current discourse tends to treat networked digital media in a vacuum, as if the peer-to-peer campaigns cropping up on blogs and social networking sites are entirely native to the digital era and have no social and cultural precedents beyond the advent of the World Wide Web. As we saw the previous section, it is true that the Internet has greatly expanded the expressive capabilities of media nonprofessionals while lowering the costs of participation—allowing citizens to forward detailed journalistic information as well as brief slogans, and video as well as still images and graphics, with remarkable quickness and ease. It has also greatly blurred the boundaries between political expression and activities associated with more traditional forms of organized activism, such as protest mobilization and the petitioning of public figures. However, much of what fuels the contemporary slacktivism debate involves social and cultural practices that not only predate the Internet, but have also been developing and growing for decades, if not centuries.

At its core, the slacktivism debate is not about whether Internet technology itself is either good or bad for political participation and the performance of citizenship. Rather, it is but the latest iteration of a much broader controversy about the value of citizen-level symbolic action as a strategy for influencing publics and, in turn, shaping the political and social world. In other words, it is largely a debate about the fundamental value of the citizen marketer approach, which, as we have seen in this chapter, has a long, rich, and complex history. Although the Internet has afforded peer-to-peer political promotion a new level of public prominence—as well as inflated levels of both hype and cynicism—these practices have their roots in everything from banner-waving public demonstrations of the 19th century to the slogan-covered T-shirts and buttons of the postwar youth counterculture. The winding narrative presented in this chapter is an attempt to historicize the citizen marketer approach in a way that moves beyond a rudimentary technological determinism, to uncover the broader social and cultural impulses that inspire many of today's citizens to boldly announce their political opinions to their peers in a variety of expressive venues.

As we have seen, the desire to bear the symbols, flaunt the colors, and forward the slogans of one's chosen political factions is perhaps as old as politics itself, yet began to take on a particular significance with the rise of capitalist liberal democracies that adjudicate power in a market-like model of contesting public appeals. The tradition of participatory collective assembly and spectacle that matured in the 19th century prior to the advent of electronic mass media brought these citizen-level practices to the center of electoral promotion and social movement advocacy—and this tradition remains strong, if somewhat more peripheral, today. However, it is the consumer culture of the 20th century, with its charge of constant self-expression in all aspects of life, which has fueled the resurgence of citizen marketing practice that continues to grow and intensify in the contemporary digital era. First modeled by those who used the accoutrements of personal style to mark their rebellious difference from the cultural mainstream, the expressive capabilities of consumer culture have since been appropriated by individuals and groups from across the political spectrum as a means of promoting their preferences and allegiances in the casual spaces of everyday life.

But what exactly marks the perceived persuasive efficacy of showing off one's politics to one's peers? What specific rationales motivate those who engage in these practices, and what do they hope to accomplish? By shifting the discussion from the specific technologies and tactics of citizen-level political promotion to the broader functions they intend to serve, we can gain a more robust understanding of these practices in the social media age and beyond. Rather than examine each new trend or platform in isolation, we must look more closely and carefully at the common logics of citizen marketing that bridge online and offline venues of participation, and that cut across the formal messaging of organized campaigns and the informal expressions of popular culture. The following chapters work to initiate this project by exploring three broadly defined logics that guide citizen marketing practice in a wide variety of contexts: making political identities eminently visible as a means of shifting social norms, dramatizing electoral momentum as a means of shifting voter enthusiasm, and raising awareness about select issues as a means of shifting public agendas. In each case, the persuasive goals of peer-to-peer marketing activity provide a unifying framework of analysis, although, as we will see, the ways in which citizens negotiate their roles as microlevel agents of political persuasion are sometimes complex and fraught with tension.

As we now turn to the testimony of citizens who selectively forward political messages to their peers, the expansive perspective of this chapter's historical lineage serves as a guidepost. In particular, the linkages that we have drawn between networked digital media and expressive objects of material culture, such as slogan bumper stickers and T-shirts, are woven into the fabric of the following chapters. Furthermore, the contexts of formally organized political assembly, such as the rally, protest, and parade, are considered alongside more casual, everyday contexts in which political expression is intermixed with popular entertainment and cultural self-expression.

The reason for tying these threads together in the current moment is that they are all indeed occurring simultaneously and are complexly intertwined. Rather than disappearing from the political landscape, public demonstrations and collective assemblies have been reinvigorated by peer-to-peer mobilization and promotion via social media. Rather than transferring wholly from offline "physical" space to online "virtual" space, the bearing of political slogans and logos takes on new hybrid forms as objects such as handheld signs and banners are remediated via digital photography and video, and as political memes become political T-shirts and vice versa. Today's participatory political promotion efforts are truly hybrid and transmedia affairs, encompassing both organized rallies and informal personal displays, both online and offline sites of circulation, and both information and symbols. Whether it is under the auspices of advocacy campaigns for social issues, election campaigns for candidates, or truly any political movement of recent vintage, we can locate citizen marketing practices in a wide variety of interconnected platforms and contexts. It is with this expansive purview that we begin our investigation of citizen marketing's overarching logics.

3

Self-Labeled and Visible Identities

Wearing Your Politics on Your (Media) Sleeve

If you logged on to Facebook in late March 2013 and scanned your newsfeed, you were likely to witness a sea of red-colored profile pictures, each with a large equal sign at the center. Some were made up of two simple pink bars, whereas others were in the shape of bacon strips, hot dogs, and dog bones. Still more equal signs were juxtaposed with pop culture icons from *Sesame Street, Peanuts,* and *Super Mario Brothers* and with the corporate logos of Orbitz and Bud Light.[1] Yet despite the many variations of this Internet meme, the red profile pictures were largely unified around putting forth a single political message: the demonstration of widespread public support for the marriage rights of gays and lesbians as the U.S. Supreme Court began its hearings on the Defense of Marriage Act and Proposition 8. For those on Facebook that week, it was an image that was nearly impossible to ignore.

This outpouring of political expression on social media, masterminded by the marketing department of the LGBT advocacy group Human Rights Campaign (HRC), enjoyed an immense level of citizen participation. Facebook would eventually estimate that as many as 2.7 million people changed their profile pictures on the first day of the campaign alone,[2] spreading the red equal sign's pro-gay message to a plethora of friends and followers. In the words of Anastasia Khoo, the marketing director for the HRC, the red equal sign profile picture campaign offered "an easy way for people to feel involved" with the issue of same-sex marriage equality.[3] The description is similarly apt for the organization's call to wear red clothing during the Supreme Court hearings as a way of further showing off one's support for the rights of gay and lesbian couples. On the HRC website, one could even purchase red equal sign T-shirts to display one's identity as a marriage equality supporter in physical space as a parallel to the profile picture's circulation in digital space.

Khoo's remarks, however, beg the question: are such ostensibly easy and low-cost ways of *feeling* involved in political issues by displaying one's allegiance also providing citizens with ways of actually *being* involved? In the press and blogo-sphere, the red equal sign campaign touched off yet another round in the public debate over what one commentator described as the act of making one's politics "immediately, graphically, demonstrably obvious"[4] via one's personalized media platforms. One popular blog piece dismissed the profile pictures as "just another form of passive activism that isn't advancing the cause" and beckoned marriage equality supporters to "do some actual work . . . instead of downloading an image and clicking a few buttons."[5] On the other end of the spectrum, one vocal sup-porter of the campaign pushed back against this familiar charge of slacktivism by remarking in the press that "with each avatar, each celebrity, each Tumblr showing support for gay marriage, public opinion began to turn . . . and I'm of the opinion that each avatar helps push the overall public opinion in the right direction."[6]

This notion of incrementally transforming others' viewpoints by visually demonstrating a critical mass of those who belong to a particular social or political group is one of the key logics that account for how citizens might potentially impact the world around them by participating in the spread of media content. Although it may strike some as overly naive to think that sim-ply announcing one's political identity with symbolic artifacts can amount to an effective form of political action, such an approach appears to grow more and more popular with each call to "show your support" through the myriad expressive channels of participatory culture. The following chapter explores how these tactics of public visibility via peer-to-peer platforms inform the citizen marketer approach, with particular attention to the identity politics of social movements. After discussing some of the important theoretical roots of this concept—in particular, Hannah Arendt's notion of the public realm of appearance—the chapter turns to an extended case study of participatory vis-ibility campaigns in the gay and lesbian movement. For decades, "coming out" in public with the help of symbolic visual markers has been a central politi-cal project for members of the gay and lesbian community, in both organized and casual everyday contexts. It therefore provides an archetypical example of how participation in public projects of visibility has been conceptualized as a meaningful—and potentially world-changing—political act. However, coming out as a political tactic is no longer limited to gay and lesbian citi-zens (if it ever was to begin with), and the chapter also addresses how the act of making identities publicly visible is becoming a core logic of persuasion and promotion for a wide range of political constituencies, from Occupy Wall Street activists to young conservatives. At the same time, the long-standing debate within the lesbian, gay, bisexual, and transgender (LGBT) movement about the potential pitfalls of visibility as a political strategy helps illuminate some of the crucial issues at stake when political identities are conflated with branded organizational logos and other packaged symbolic markers. As we will explore, such a strategy provides opportunities for myriad identity groups to publicize their existence and potentially shore up support by shifting social norms and altering perceptions of political reality, yet it also risks flattening diversity within these groups as certain categories and characteristics become privileged over others, potentially reducing political identity to the most easily marketable formulas.

Self-Labeling in the Public Realm of Appearance: From "Walking Billboards" to Visible Identities

In her landmark critique of modern branding strategy, Naomi Klein draws attention to how displayable commodities such as clothes have been increasingly inundated with brand logos as a way of putting consumers to work as word-of-mouth or viral marketers in public spaces. As an example, she quotes a media executive who describes the launch of a branded men's clothing line that is intended to create "a nation of walking billboards" to promote his company.[7] This term, familiar in popular discourse, signals the conventional way in which the personalized display of promotional content has often been conceived. Typically carrying a derogatory inflection, "walking billboards" or "walking advertisements" paints a picture of participants in peer-to-peer promotion as veritable empty vessels, giving themselves over to the brand as they become mindless links in a chain of strategic public persuasion. For instance, in the documentary *Crumb*, the underground cartoonist R. Crumb goes on a memorable rant about his disgust with the branded apparel trend: "Everybody walks around, they're walking advertisements . . . Walking around with 'Adidas' written across their chests, '49ers' on their hats. Jesus. It's pathetic. It's pitiful. The whole cultures' one unified field of bought–sold–market researched everything, you know."[8]

Why, one may ask, would people voluntarily serve as walking billboards? Perhaps they are simply not cognizant of their role in promotional culture and are just mindless dupes. More likely, however, they gain something of value from the process, particularly in terms of making a statement about their own identities. Indeed, whether selectively forwarding promotional content via their chests, their car bumpers, or their social media profiles, so-called brand evangelists or advocates forge a dialectical relationship between their public identities and whatever it is that they are helping to promote. Using their personal venues of self-expression to relay another's promotional message, they simultaneously send a message about who they are as individuals that may benefit their social standing (e.g., "I'm a Tommy Hilfiger kind of guy."). Commercial marketers have long sought to exploit this relationship for strategic gain, associating brands with certain aspirational identities and lifestyles to encourage consumers to see themselves—and even more important, to outwardly express themselves—as members of a brand community.[9]

For citizens who voluntarily act as grassroots intermediaries for political brands, broadly speaking, this dynamic of identification and personal representation appears to be even more crucial. In part, this can be linked to the phenomenon whereby some individuals choose to publicly align themselves with political campaigns and activist causes to craft a socially desirable image for personal gain (i.e., what has been derisively labeled in the press as "virtue signaling"[10]). It is certainly likely that some who post content on their social media profiles showing support for a cause do so out of a desire to look more caring, compassionate, or worldly. Likewise, some who flash political logos and slogans on their T-shirts or bumper stickers may be motivated by an interest in appearing socially conscious and responsible to their peers. In the context of political expression on social media, there is some empirical data to support this idea. For instance, a study of Israeli youth found that many who posted about the 2014 Israel–Palestine conflict did so to "create an impression of themselves as someone who is knowledgeable and opinionated around politics and to receive positive social feedback for having candidly expressed their views."[11]

Although the authors of the study frame this as a relatively positive social outcome, other scholars have pointed to such self-interested motivations for expressing political identity online as raising key concerns. Dahlgren, for example, refers to the pattern in social media–based political expression of "personalized visibility, which includes self-promotion and self-revelation" as inherently worrisome[12] and connects it to a consumer-like mode of political participation that privileges personal satisfaction at the expense of broader democratic engagement. Papacharissi makes a similar point in her analysis of what she calls the "private sphere" of online self-expression, in which the technological structure of social media profiles and feeds "invites expression, affords autonomy, and enables control of the self and its multiple performances."[13] As she argues, these dynamics "present ego-centered needs and reflect practices structured around the self. This suggests liberating practices for the user, but not necessarily democratizing practices for the greater society."[14] In other words, Papacharissi suggests that online political self-expression may be valuable for developing one's own political identity, yet it contributes little in and of itself to the broader political discourse.

However, although personal identity construction and public image maintenance may be important factors in accounting for why some choose to show off their political attachments to their peers in both online and offline venues, such a framework does not consider how these self-identifying practices can also serve as means of persuasion. When I began interviewing citizens who express their political views in public through various mediated channels, I was struck by how often they equated the act of making a statement about who they are as individuals with sending a political message that was deliberately intended to influence the viewpoints of their peers. Crucially, the persuasive force of these messages was perceived to have less to do with the specific content being disseminated than with what the act of *identifying* with it might represent in the minds of others. Thus, the notion of the walking billboard as simply a conduit or relay for promotional material fails to capture the full complexity of how political self-expression is employed in the citizen marketer approach. For those who willfully participate in peer-to-peer political promotion, persuasion is not only conceptualized in terms of giving further public exposure to a preexisting message. Rather, they often imagine their own identities as contributing a core aspect of the overall influential message that they wish to send. In such scenarios, the forwarded media content functions as a self-affixed label to publicly announce a key attribute of one's identity that would not have been obvious or visible otherwise, thus making a politically charged point about its presence in the social world.

To be certain, some salient aspects of identity are already visibly marked on the body without the need for symbolic labels. As Linda Martín Alcoff points out in her book *Visible Identities*, categories such as race and gender have become fundamental for identity politics not because they are inherently natural or biological, but because they have been continually imposed on bodies through social and historical patterns of perceptual practice that have real-life consequences for how these bodies are differentially treated by others.[15] However, although the perceptible characteristics of the human body play a significant role in many of the visual practices considered here, my focus is primarily on how individuals and groups draw on the semiotic universe of signs and images to make legible aspects of identity that lay outside the realm of "common-sense" perception. As we will see, by making previously illegible identities eminently visible through conspicuous forms of self-labeling, one may see him- or herself as strategically challenging public perceptions of who exactly "the people" out there really are. The message being sent is thus one that attempts to implicitly redefine the nature of social reality for the audience, as opposed to one that attempts to explicitly persuade it in one direction or another. However, reorienting perceptions about who the people are is an act with significant—albeit indeterminate—political consequences, since notions of what *is* precede notions of what should and should not be. Thus, rather than being limited to only self-serving impulses, the expression of personal political identity may take on a far more public-minded character.

In the field of political philosophy, the act of marginalized groups making themselves visible and legible in public—informing both one another and outsiders of the fact of their existence—has long been recognized as politically potent. For Hannah Arendt, self-disclosing appearance in the public realm serves as the basis for all political action. In her landmark work *The Human Condition*, Arendt relates a trenchant anecdote from ancient Rome to illustrate this point, in which a proposed law to publicly identify slaves by mandating a dress uniform was rejected by Roman authorities: "The proposition was turned down as too dangerous, since the slaves would now be able to recognize each other and become aware of their potential power. . . . What the sound political instinct of the Romans judged to be dangerous was appearance as such."[16] Arendt also describes how the visually marked out appearance of the *sans-culottes*—the working-class partisans of the French Revolution who wore distinctive garb to set themselves apart from their opponents—coincided with their being "admitted into the public realm" and "enter[ing] the scene of history."[17] Arendt's conception of self-disclosing appearance thus suggests the political force of visibility tactics: by making identities and shared viewpoints publicly perceptible through symbolic markers, the collective can potentially gain in strength and numbers while confronting the broader public to become attentive to its demands.

In recent years, Arendt's ideas about the public realm of appearance have proven inspirational to scholars who are interested in charting the alternative communicative practices of social movements that often defy traditional norms of public advocacy. For instance, the queer theorist Michael Warner celebrates "the scene of world making and self-disclosure" suggested by Arendt's model as opening up possibilities for "visceral" modes of collective expression among groups committed to social change.[18] For Warner, this understanding of political participation as including symbolic expressivity (as well as more ostensibly rationalist modes of public dialogue) is particularly salient for the gay and lesbian community, which has historically been excluded from the idealized public sphere of deliberation.[19] Warner is primarily interested in how such expressive practices help give internal shape to a queer public—conceptualized as a counterpublic because of its emphasis on social transformation. However, we can recognize how the visual appearance of new publics and counterpublics on the "scene of history" can also provide a means for these groups to speak to and act on the broader external society in visceral terms. Andrea Brighenti describes this distinction as one between intervisibility, in which members of a group become visible to each other in ways that build community and lay the groundwork for mobilization, and public visibility, in which these group members can appear in public settings and participate in public discourse.[20] It is this dual, overlapping role of internal and external visibility that forms the core of this chapter's discussion.

Regarding the latter, how exactly would citizen-level acts of visual self-disclosure take on a persuasive character in the broader public landscape? One possible clue is suggested by Lyman Chaffee in his global comparative study of protest street art. Conceptualizing street art as "low-technology mass communication in an age of high technology"[21] (including wall murals, graffiti, posters, signs, stickers, pins, and T-shirts). Chaffee argues that its personalized and intimate form of presentation carries certain rhetorical connotations of authentic grassroots political sentiment, standing as signposts of what is "really" going on in the streets: "its use is appealing especially to those who stress a collective consciousness and claim to speak for and represent the people."[22] Here, Chaffee recalls the ideas of the political theorist Ernesto Laclau, who posits that laying claim to representing the people by articulating who "they" are and what they believe is central to the winning of power by both hegemonic and subaltern groups engaged in political struggle.[23] The personalized protest street art that Chaffee documents has readily apparent connotations of authentically representing the people in this fashion, as it seemingly emerges spontaneously from the grassroots rather than from top-down systems of communication. When popular, participatory forms of expression are combined with a living, walking body (as in the case of T-shirts, buttons, or pins) or even with a digital avatar (as in the case of social media profiles), this appeal to grassroots authenticity is particularly charged. In other words, the political message being put forth has a palpable concreteness to it, coming from a real person and not from a disembodied, professionalized elite. Such an authenticating appeal may be even more pronounced when personalized self-expression is coordinated into collective spectacles, articulating an aggregate image of the people that can effectively challenge common-sense perceptions about the makeup of social reality and, in turn, reorient social norms. Thus, the formal qualities of these expressive public displays—particularly having to do with the intimate, personalized, and embodied ways in which they are presented—make them particularly appealing for those claiming authentic representation of the people for persuasive political ends.

Participatory Visibility in the Gay and Lesbian Movement

For the gay and lesbian movement that has been gaining political momentum since the 1960s, projects of public visibility have become particularly salient in efforts for social transformation. As the queer theorists Lauren Berlant and Elizabeth Freeman have argued, "visibility is critical if a safe public existence is to be forged" for members of sexual minorities that have historically faced widespread discrimination, oppression, and silencing.[24] Indeed, cultural invisibility has long been seen as a primary obstacle to social and political progress for members of the gay and lesbian community. One popular solution has been to expand cultural representation in the mass media, including both journalism and entertainment.[25] However, more personalized, grassroots-level expressions of identity have also been used to achieve the goal of gay and lesbian public visibility, making literal the movement slogan "we are everywhere."

For instance, some gay activists concerned with expanding social and political recognition have created organized spectacles of same-sex intimacy in public. Charles Morris and John Sloop describe public performances such as "kiss-ins" as creating a "representational critical mass" of queer bodies that demonstrates the presence and power of this group to the broader society.[26] Berlant and Freeman further describe how these organized kiss-ins, often staged in shopping malls and other popular public spaces, have been used by activists to "broadcast the ordinariness of the queer body" and "create public spaces that are safe for the visible manifestations of multiple sexualities."[27] In a similar vein, Kevin DeLuca examines how the activist group ACT UP performed theatrical public protest actions such as "die-ins" to challenge the societal taboo on the visible homosexual body: "in making their bodies visible, present, and exposed, the ACT UP activists call on society to care."[28] This idea, which recalls Arendt's point about the power of appearance among the French *sans-culottes*, effectively encapsulates the rhetoric of participatory visibility that gay and lesbian citizens and their allies have forged through a variety of expressive platforms. Furthermore, efforts like the ACT UP demonstrations underline how acts of political self-disclosure are not inherently confined to a self-serving private sphere of personal expression; rather, they have been effectively mobilized by numerous political organizations to create coordinated forms of public advocacy. In the digital age, however, these kinds of coordinated visibility campaigns seem to be transitioning from a logic of collective action to a logic of connective action, in which expression becomes more atomized as it is performed via individual social media profiles and feeds.[29] As we will see, this shift carries with it important possibilities for diverse and critical voices to emerge within the spaces of political visibility projects, which, in turn, may weaken the ability of organizational entities to control the images they seek to make visible to the broader public.

In recent years, critics within the gay and lesbian movement have called traditional visibility strategies into question precisely because of their seemingly limiting and managed qualities. In particular, there has been pushback against the celebration of cultural visibility in the mass media, which often places undue restrictions on how the community is represented to the wider society. Suzanna Walters, for instance, argues that the quality of media visibility is far more important than the quantity and that many of the images of gays and lesbians that appear in contemporary television and film promulgate reductionist and even harmful stereotypes.[30] Furthermore, a range of critical feminist scholars have questioned the value of identity politics more broadly as an effective framework for challenging multiple and interconnected forms of social oppression, of which heterosexism is only one part. Tina Chanter, for instance, argues that the visibility efforts of identity politics tend to privilege one abstract category of identity to the exclusion of all others, which not only discounts important differences within these categories, but also reproduces dominant social structures: "If a group of Western women organizes around the identity 'women,' that identity tends by default to privilege middle class, white, and heterosexual identity, and to erase the significance of experiences of working class, raced, and gay women as non-normative."[31] It would seem that more participatory forms of mediated visibility would help mitigate such concerns, potentially allowing for the full diversity of social subjects to enter the public realm of appearance rather than just a restricted set of palatable or privileged representations.

However, the issue remains that a great amount of peer-to-peer political expression at the citizen level relies on the selective forwarding of professionally produced content—in some cases created by the marketing teams of major advocacy organizations—that may place undue limitations on the representation of identity. In the gay and lesbian movement, popular icons such as the HRC's equal sign logo and the rainbow pride flag have formed a standardized symbolic repertoire for its members to express their identities and contribute to aggregate projects of visibility. Yet at what cost? How might political and social identities be constricted and flattened by being placed under packaged symbolic banners, and what are the consequences? At the same time, how might the expressive fluidity and malleability of digital media enable citizens to actively challenge this process and move beyond the constrictions of a standardized visibility politics?

Before we tackle these crucial questions, however, it is first necessary to take a closer look at how the practice of making identities visible through symbolic self-labeling plays out within everyday lived contexts. It is here that we turn to the voices of citizens who use a variety of media platforms to strategically announce their gay, lesbian, and ally identities on the scene of history.

VISIBILITY FOR EXTERNAL AUDIENCES: CHALLENGING PERCEPTIONS OF THE PEOPLE

As noted previously, one of the most popular campaigns yet to use the participatory self-labeling tactic involved the coordinated changing of Facebook profile pictures to a red version of the HRC's logo during the 2013 Supreme

Court hearings on same-sex marriage. When interviewing Facebook users who posted the red equal sign images about what they hoped to achieve, the theme of visibility as norm-shifting political persuasion came up time and time again. In other words, their rationales for how the campaign worked rhetorically tended to focus less on the specifics of the logo itself than on the act of adopting it as a visual manifestation of their own identities. By connecting their digital avatars on Facebook to the image, participants in the campaign made themselves publicly legible as marriage equality supporters, and many respondents pointed to the self-disclosing nature of these acts as key to the campaign's overall persuasive power.

For instance, as Ron explained, "I decided that if I did it, the people that saw my posting would be more aware of the enormity of those that were in favor of it, if not gay also." Ron felt that the value of the campaign was "showing that more people were in favor of it than against it, that there was a sizable number. . . . The influence I thought just would be making people more aware of the numbers of us that were in favor of it or were affected by it. And maybe change their way of looking at it, and show that it's not a minority kind of fringe group." This same logic was also voiced by Chloe, who described that her goal in posting in the red equal sign was to "demonstrate to others how overwhelming, you know, with the volume of people that changed their profile pictures, how overwhelming in support in particular this sort of Facebook generation is on this issue . . . when reasonable people who are against gay marriage see how many reasonable people are for it, I think that that changes their minds slowly over time." According to Ron and Chloe's line of reasoning, the collective visual representation of gay and lesbian–supportive identities has the potential to alter public perceptions of political reality in such a way as to compel others to follow suit, creating a kind of bandwagon effect. By forming a representational critical mass on Facebook (to borrow Morris and Sloop's term[32]), the hope of these citizens is that the full panorama of gay rights supporters might come into focus and be recognized and considered by the broader society.

In the field of marketing psychology, this concept is sometimes discussed in terms of setting descriptive norms for others to follow, as opposed to prescriptive norms that directly tell others what to do or think.[33] Writing for *Scientific American*, the social psychologist Melanie Tannenbaum draws a parallel between this bandwagon-creating dynamic of the red equal sign campaign and the setting of descriptive norms as an effective means of political persuasion. As she explains, "based on everything that we know about our brains and their bafflingly strong desires to fit in with the crowd, the best way to convince people that they should care about an issue . . . isn't to tell people what they should do— *it's to tell them what other people actually do*" (italics in original). Tannenbaum thus offers a plausible explanation for how articulating an image of the people may hold a powerful, if indirect, influence over public opinion. In the case of personalized political expressions such as social media profile pictures, we can extend Tannenbaum's point and suggest that for many who engage in citizen marketing practices, convincing people is a matter of *telling them who other people actually are*. By describing the makeup of the social and political world in ways that challenge conventional wisdom, these citizens hope that those who wish to "fit in with the crowd" will be compelled to reconsider their positions.

Such a tactic can be found on the grounds of many college campuses across the United States, where T-shirt campaigns are frequently used to announce the presence of gay and lesbian students to their classmates (often in conjunction with events such as National Coming Out Day). These coordinated efforts are typically organized by campus LGBT and queer alliance groups via mass email or social media posts, and the T-shirts are often created and distributed specifically for these events[34]—again, we note the role of organizations in spurring individual acts of political self-disclosure. On the chosen day, the bodies of the T-shirt wearers deliberately work in conjunction to produce a collective spectacle announcing the widespread presence of gay and lesbian identities, as well as those who take on the identity of allies of the community. Brian, a campus leader at his university, emphasized the value of using these sorts of T-shirts to promote the gay and lesbian cause on his campus, explaining that the point of such activity is to "make the non-LGBT world more aware of the fact that there is an LGBT world and it's pretty strong at our school, and we're the same as them and we should all come together and have the same rights." By changing perceptions about how many gays and lesbians are in fact members of this campus community, Brian suggested that they could influence others in the community to rethink their attitudes toward this group. In other words, once the student body knows that gays and lesbians are all around them, it will likely become more sympathetic to their concerns.

However, this strategy of making identities visible for the purpose of political persuasion need not always rely on a mass-coordinated articulation of the people—that is, dramatically showing scale and numbers as a sign of strength. For instance, Jennifer noted that she imagined the individual persuasive power of making her lesbian identity visible by wearing the popular "Legalize Gay" T-shirt in public settings such as grocery stores:

> If we want to get rights, people need to know who those rights are going to, because once you put a face to the cause it makes it a little bit harder to say that you don't care. When people say that they don't agree with gay rights, it's much easier for them when they don't actually see a human being who's being denied those rights. With me wearing [the T-shirt], it's very clear who I am.

As Jennifer suggested, her body becomes a persuasive message for gay and lesbian rights when she publicly labels her sexual identity with a symbolic visual marker. By standing as a humanizing representation of the sexual minorities in her local community, she forces all those who walk past her to consider the fact that this political struggle affects living, breathing people in the vicinity such as herself.

This theme of humanizing the issue through personalized and embodied appeal was also common among participants in the digital context of the red equal sign profile picture campaign. Whereas some suggested a kind of diffuse bandwagon effect produced by a wide-scale articulation of identity online, others emphasized the more specific capacity of individuals within a person's social network to lead by example. Here, the fact that social media platforms like Facebook connect people to their own friends whom they admire and trust—and who may thus influence them at an interpersonal level, as suggested by Katz's and Lazarsfeld's classic theory of two-step flow[35]—becomes integral for how social media–based visibility campaigns are conceptualized as effective political persuasion. For example, Sarah, who self-identified as a straight ally, explained that she hoped that those who are against marriage equality would react to the sight of their friends' red equal sign profile pictures by thinking, " 'well maybe if all these people I respect support this thing, maybe I should rethink what's going on here. Maybe I'm not making the right decision here.' " Peter, also a straight ally, further elaborated on this idea: "If you have someone you really admire, and you don't know what they think about gay marriage, and you see them with a red equal sign . . . I think it can totally shift perceptions. Who our friends are affect us, our behavior, our opinions."

Along similar lines, some who saw their participation in the red equal sign campaign as announcing their own gay or lesbian identities emphasized the importance of personal connections with Facebook friends who may not yet support marriage equality rights. As Jonathan put it,

> I hope that my friends who are straight, or who don't really agree with me politically, would look at [the profile picture] and say "yeah, that [Supreme Court] decision affects Jonathan" . . . Saying "this affects me" makes it a lot harder for somebody to get up and say "that's wrong, that should never happen." It changes the conversation and it changes the tone, if that same person says "that shouldn't happen except for Jonathan," or except for somebody else that they know who it's directly affecting.

For some, this effort was reinforced by interactions that users were able to have with their Facebook connections as a result of posting the profile picture. David noted that he used his red equal sign picture to initiate dialogue regarding how the marriage equality issue affects him personally:

> The bulk of my friends are straight people who didn't have any clue what marriage equality was about and why it was important. So there were a lot of questions about "hey, what is the big deal?" And I was able to point out the benefits of marriage that married people never think about. . . . I was able to say, you know, here are the things that you get because you're married to your wife that I don't get in most states being married to my husband.

Thus, David was able to use his act of mediated self-labeling as a gateway for more in-depth rhetorical exchanges, with the profile picture serving to visually establish the argument "this affects me."

For Mason, this sort of visibility achieved via Facebook may have a deeper political impact than mass media representations of gays and lesbians because of the platform's interpersonal nature. As he explained regarding his motivations for participating in the campaign, "I wanted to make sure that real life representation was also there." Although he acknowledged the value of celebrities coming out publicly in the media, he remarked that "I don't think it's touching people on a normal level though. . . . When you actually have somebody that says 'you know what, I know you, we go out for coffee, and I'm gay, too,' that kind of changes it, makes it personal. . . . For a lot of people, they need to know that there are real people out there." Although such a strategy of coming out to straight peers has been a focus of the organized gay rights movement for decades, it appears that social networking sites and other participatory media platforms are multiplying the ways in which gay and lesbian citizens can coherently and explicitly self-disclose their identities for the purpose of peer influence.

Furthermore, although the red equal sign campaign presented an opportunity for some to advance the project of gay and lesbian public visibility via digital means, some respondents emphasized the particular persuasive significance of coming out as straight allies. Melanie suggested that her heterosexual identity, coupled with the red equal sign image, may have boosted the rhetorical power of her own participation in the campaign: "As, you know, a straight married heterosexual couple with a family . . . I think that might surprise people, and that might surprise some of my more right-wing kind of Facebook friends. And I just thought, well, if they take note it's because they kind of know who I am."

As these varying accounts suggest, the visibility rhetoric of efforts like the red equal sign campaign is complex and multifaceted, with each new participant potentially adding layers of meaning by juxtaposing the standardized political symbol with the specificity of his or her own identity and social position. This dovetails with a broader point made by Henry Jenkins regarding the importance of individual Internet users in "shaping the circulation of media content, often expanding potential meanings" and "allowing media content to be localized to diverse contexts of use."[36] In other words, each time an individual adopts a symbolic identity label via his or her profile picture, T-shirt, or other personalized platform of expression, he or she creates a set of political meanings that are unique to his or her own social context. Thus, in addition to enabling collective articulations of the people for the purpose of broad-scale norm-shifting, participatory media also allow for visibility rhetoric at the more granular level of personal influence, reaching the public one person—or Facebook friend—at a time.

VISIBILITY FOR INTERNAL AUDIENCES: SHORING UP COMMUNITY AND PRIDE

Despite this emphasis on changing the minds of those who do not yet support marriage equality, a potentially oppositional external public was not the only audience that was targeted by Facebook users who posted the red equal sign as their profile picture. According to Anastasia Khoo, the director of marketing for the HRC and creator of the campaign, "we wanted to send a message of hope to young people who might not feel accepted, or realize that there is a visible community of people out there who are supportive. That's when I had an idea for a campaign that could harness the power of Facebook."[37] In keeping with this theme, some respondents saw one of the goals of the campaign as heightening a sense of social support for those who may be struggling with their sexuality, thus lifting their morale and self-esteem. In this sense, the campaign was framed as a kind of public mental health outreach effort for at-risk LGBT youth, albeit one that also has strong political overtones because of the embattled status of this group within the broader society. As Mia commented regarding the campaign,

> children who thought that they were alone, you know. . . . Here's one gay high school student thinking that he's the only gay person in the world and he's got no idea of who to look up to. He can go online on Facebook, do a search for red equal sign and can see all the different organizations that are affiliated, that represent themselves behind that sign. And so it really gives them the ability to not feel alone anymore.

In a similar fashion, Mason noted that he participated in the campaign in part because

I thought that there's a lot of kids out there that are on Facebook, probably behind their parents back, and they need to know that they're not
alone, especially with the entire wave of suicides. . . . You need to know
that you have a community base, you need to know that there are people who support you and will encourage you when you need it.

Furthermore, Mason noted having conversations with gay and lesbian youth
on Facebook along these lines in direct response to his red equal sign profile
picture, emphasizing his success in helping these community members improve
their attitudes toward their own sexual identities.

This model of a public-facing visibility campaign doubling as an intracommunity morale-boosting effort has become common in the contemporary gay and
lesbian movement. One particularly popular (and fairly controversial) example
is the "It Gets Better" online video project founded by the gay rights activists Dan
Savage and Terry Miller. The videos, which became a YouTube hit in 2010, feature gay and lesbian individuals discussing their first-person childhood experiences of oppression and bullying, along with how their circumstances improved
later in life—a message that has been both lauded as inspirational and lambasted
as deceptively over-optimistic. As the organization's website explains, the goal of
its creators is to "inspire hope for young people facing harassment. In response
to a number of students taking their own lives after being bullied in school,
they wanted to create a personal way for supporters everywhere to tell LGBT
youth that, yes, it does indeed get better."[38] Spurring participation from myriad
"out" gay and lesbian celebrities as well as from supportive public figures such
as President Obama, "It Gets Better" centered on the idea of mediated public
visibility as a pride-boosting antidote to the problem of LGBT youth depression
and suicide. The majority of the participants, however, were everyday Internet
users rather than public figures, and the project eventually included more than
50,000 user-created videos. One of these participants, a young gay man named
Calvin Stowell who experienced intensive bullying as a child, remarked in *The
New York Times* that "growing up, I never had someone to confide in . . . I can't
even articulate how much this has ended up meaning to me," further noting that
over a dozen teenagers experiencing suicidal feelings had emailed him as a result
of his video.[39] Like the red equal sign campaign, the "It Gets Better" project fashioned an online representational critical mass that targeted at-risk members of
the community while also coming into contact with a networked external public.

The potential intracommunity benefits of participatory visibility projects
have also been a central focus of the T-shirt-based campaigns on U.S. college
campuses. As discussed by students who participate in these campaigns, a
major goal of this mediated self-disclosure is to persuade others within the community's gay and lesbian population to be more out and proud. For instance,

Jessica gave this account of a campus-wide event that she participated in at her college where students wore T-shirts reading "Gay? That's Okay": "The idea was to get as many people as possible to wear that shirt on the specific day to just show the support visually, and to have those people who are okay with it being out in the open show it. For those people who aren't okay with it, when they see all those people that are okay with it, it might incline them to become more comfortable in who they are." Brian offered a similar account of an event on his campus in which students wore matching shirts simply reading "Supporter": "A lot of people in the closet don't feel like they have support, and they're really unaware of all the people there that do support them . . . so it's an important message to get across." He went on to explain that wearing the T-shirts helped to reinforce a shared sense of identity among LGBT students on his campus: "It's just showing I'm a part of something, and that I'm not alone, I'm part of a greater thing. It kind of builds community when you wear the shirt and you see someone else also wearing the same shirt. That builds a sense of solidarity I guess."

Brian's point about community building draws attention to how such inward-facing visibility campaigns—or intervisibility, in Brighenti's terminology—have significant implications for collective political mobilization. To be sure, there are many personal and social benefits of feeling connected to any group. However, when communities such as gays and lesbians are marginalized within the broader society and when their members have been historically invisible to one another, conspicuous self-labeling can bring about a mutual recognition of the group's size and collective force that can, in turn, form the basis for collective action. This is precisely the point that Arendt makes about the dress code of the Roman slaves potentially allowing this group to become a self-aware political constituency: by coming together in the public realm of appearance and seeing themselves as a collective, they lay a foundation for organizing and asserting their political will.

Such a notion is also central to the culturalist perspective on peer-to-peer political interaction online (outlined in Chapter 1), as numerous scholars have emphasized how building a collective identity through culturally situated digital practices can serve as a preliminary step toward future mobilization. For instance, Dahlgren cites Brighenti's concept of intervisibility in his discussion of how "on the web in particular, strangers become visible to each other to various degrees in order to cooperate" in organized political contexts.[40] In the social media age, opportunities for mutual recognition and collective identity building are multiplying considerably. Writing for *Wired* magazine, Bill Wasik argues that social networking sites allow for new and potentially powerful constituencies to form as members make themselves known to one another through digitally enabled self-identification. Wasik offers the nonpolitical example of electronic music fans who have formed an enormously popular and thriving subculture in the United States, primarily through peer-to-peer online networking. As he explains, "for these groups, suddenly coalescing into a crowd feels like stepping out from the shadows, like forcing society to respect the numbers that they now know themselves to command."[41] A similar statement could be made about the gay and lesbian movement in the context of efforts such as the red equal sign profile picture campaign and the "It Gets Better" YouTube videos. Yet the Internet is only the latest medium for symbolic self-disclosure—although now certainly the most widespread and popular—that gays and lesbians and their allies have used to step out from the shadows of invisibility and forge a self-aware and empowered political constituency.

The political logic of making identities visible is thus not limited to setting descriptive norms for the broader society to follow, although such external-facing persuasion may be a significant element. Rather, it also carries an important dynamic of internal community building that may serve as a foundation for future organized activity. In the case of the gay and lesbian movement, community building begins with the social support of peers, since the enormous historical challenges that have been placed on sexual minorities make it difficult for individual members to face them alone. By coming to recognize that there are many others who are just like them, or who support them with slogans like "Gay? That's Okay," young gays and lesbians may be persuaded to feel not only a boost of personal pride but also a burgeoning sense of collective political efficacy.

In this sense, we can recognize significant overlap between the citizen marketer approach of instrumental peer persuasion and the framework of culturalist agency building via civic interactivity. Indeed, both can occur simultaneously when the same act of media spreading reaches different kinds of audiences. Moreover, we can recognize a dimension of peer persuasion in the kinds of coordinated actions that deliberately promote intervisibility as a pathway to collective mobilization. In efforts such as the HRC-initiated red equal sign campaign, influencing peers to adopt a shared sense of membership in a gay and lesbian *political* community (i.e., above and beyond a social and cultural identity group) appears to serve as a tactical means of promoting its growth more broadly, which in turn can potentially boost its fortunes on the broader political stage. In this sense, the bandwagon effect produced by norm-shifting acts of self-disclosure is seemingly targeted at internal as well as external audiences, as gay and lesbian citizens announce their presence to circulate the message, "we're here, we're queer, come join us."

Visibility beyond the Gay and Lesbian Movement: Coming Out Tactics and the Citizen Marketer Approach

The gay and lesbian movement thus offers a particularly instructive case study for understanding how the act of making identities visible through mediated self-labeling can potentially function as viral marketing–like, peer-to-peer political persuasion. However, the practice is not limited to this particular group, but is rather becoming an increasingly popular tactic for a wide swath of constituencies in an era in which political self-disclosure can be achieved through a simple button click. In fact, it could be argued that the high-profile coming out actions of the gay and lesbian movement have created a kind of model that many political groups are now following, particularly when their status in the public realm of appearance is less than fully conspicuous. Not all coming out is the same, however, and the specific risks and challenges faced by gay and lesbian citizens who publicly announce their sexual difference should not be diminished or forgotten. For instance, danah boyd points out that participating in the "It Gets Better" YouTube campaign resulted in numerous LGBT youth being at heightened risk of peer harassment and bullying.[42] Yet as platforms for announcing one's membership in virtually any group have greatly expanded with the popularization of social media profiles and feeds, coming out appears to be emerging as a core logic of the citizen marketer approach across political categories.

For instance, the trend can be observed in numerous Twitter-based hashtag campaigns that have become popular within identity-focused social movements beyond the gay and lesbian context. In 2014, for instance, a variety of feminist-oriented hashtags called on women to come out regarding their personal experiences with male-on-female violence, sexual assault, and harassment, including #YesAllWomen (inspired by a misogyny-fueled mass shooting in California)[43] and #WhyIStayed and #WhyILeft (inspired by a highly publicized domestic violence incident involving a professional athlete). Beverly Gooden, the domestic abuse survivor and writer who started the latter of the two campaigns, explained to the press that "the hashtag shows that there are people out there who have lived this, and like me, have come out of this."[44] Again, we glimpse the logic of self-labeled and visible identities as a norm-shifting strategy, in this case using the testimonies of domestic abuse survivors to put a human face on this issue and call on others to care about gender-based violence—as well as inspire other women in similar situations to both persevere and join the movement.

Also in 2014, the Black Lives Matter movement that surged in the wake of the police killing of the African American youth Michael Brown in Ferguson,

Missouri, launched a hashtag on Twitter called #IfTheyGunnedMeDown; in these posts, young blacks juxtaposed photographs of themselves wearing street-oriented fashions typically perceived as "thuggish" with other images showing them in more professional and "proper" dress, and asked rhetorically which one the media would use to represent them if they had also been slain by police. On one level, the campaign was a savvy satirical critique of biased journalistic practices, but on another level, the grassroots circulation of images of youthful African Americans in graduation robes, military uniforms, and business suits constituted an effort to shift public perceptions about race by making visible alternative images of young black identity. As the scholar Malik Houston commented regarding #IfTheyGunnedMeDown, "there is little room in the public discourse for complexity of black males and their identities . . . whether someone is wearing a hoodie or sweatpants, they can still be a good father and have a job and take care of responsibility."[45] By coming out on Twitter as responsible citizens who also sometimes dress in ways that are negatively stereotyped, participants in this viral campaign— as well as those who selectively forwarded the images on Twitter—worked to challenge racist assumptions and complicate the public image of young African Americans.

In addition to visibility efforts addressing issues of identity-based oppression that involve race and gender, coming out tactics via networked digital media have begun to appear in numerous other contexts of political activism. Wasik, for instance, draws a parallel between the above-noted visibility dynamic he witnessed in the electronic music underground and the "We are the 99%" Tumblr page associated with the Occupy Wall Street movement. For Wasik, the Tumblr page served as a digital extension of the Occupy protests, representing a "disconnected group getting connected, a mega-underground casting off its invisibility to embody itself" as a new political constituency.[46] On the "We are the 99%" Tumblr page, Americans struggling with financial security, debt, and unemployment posted photos of themselves holding up handwritten signs that explained their plight, as well as their feelings that the U.S. economic system has been unfairly stacked against them. Participants also frequently displayed the slogan "I am the 99%" on their handheld signs, thus making themselves stand as visible representations of what they perceive to be a disenfranchised and ignored economic majority.[47]

Such acts of mediated self-disclosure—cleverly hybridizing digital and corporeal modes of embodiment—signal how the tactic of coming out in public is being adopted by political movements whose concerns do not necessarily align with identity politics in the conventional sense. Even the phrase "We are the 99%" suggests how this economically focused protest movement cast itself as an as-yet-unrecognized identity group that seeks public visibility as a first step toward social and political transformation. To be certain, there are strong echoes here of traditional Marxist class consciousness-raising, and thus it would be a mistake to claim that framing economic struggle in terms of a collective identity struggle is new. Nonetheless, the "We are the 99%" Tumblr page demonstrates how contemporary political movements are utilizing the specific self-labeling tactic of participatory media visibility campaigns—so strongly familiar in the gay and lesbian community as well as in identity politics more broadly—to come together as a group and make their collective concerns known to the outside world. In fact, the political scientist Jodi Dean shrewdly describes the "We are the 99%" Tumblr page as a " 'coming out' of the closet imposed by the conceit that everyone is middle class, everyone is successful."[48]

In my own interview research, one of the most surprising places in which I heard distinct echoes of the coming out visibility tactic was among young conservatives. Although this group may not be as invisible in the broader United States, this did seem to be case in the specific Northeastern region where I conducted my research. For the young conservatives I spoke with, mostly members of Young Republican organizations on their respective college campuses, their apparent minority status within the local community compelled them to increase their public visibility through wearing self-labeling T-shirts on campus. By openly identifying themselves in this way, they hoped that like-minded others would be encouraged to step out from the shadows and follow suit. For instance, William described how he sought to become a representative model of sorts for being openly conservative on his heavily liberal campus:

> I think in this current political climate, people would be afraid to sign up for [College Republicans] because they think people are going to hate them . . . that because they're conservative, they're somehow a bad person. So it's kind of like me going out there and being like "yeah, I'm proud to be a conservative and I'm proud to talk about it." If I'm out there doing it, people who maybe were nervous or didn't want to outwardly support a candidate in fear of retribution, or at least being thought of in a different light, might think "hey, if there's other people doing it then maybe it's okay for me to as well."

Thus, William posited that publicly showing off his minority political identity within a potentially hostile environment could serve as an influential model for other young conservatives to follow. Somewhat ironically, his comments strongly recall the pride-boosting visibility campaigns of his gay and lesbian classmates, albeit in a different political register.

Brandon provided a similar account regarding his motivations for wearing a conservative Tea Party–oriented T-shirt at a large televised Democratic rally near his campus that featured an appearance by President Obama. As he explained,

> going there and getting on TV lets everybody who is watching realize that not all young people age 18 to 24 are Obama supporters. There are some people that dislike Obama as much as others like him, and I think that needs to be put out there . . . because there are a lot of people that are my age that aren't happy with what's going on, but they just don't really have a voice.

Here, Brandon framed his televised act of self-labeling as visually announcing the existence of young anti-Obama conservatives to a mass audience who might be otherwise unaware. Furthermore, he suggested not only that the politically like-minded would be encouraged by this public display of conservative identity to feel not so alone and invisible, but also that such an image could challenge the broader public to rethink who young Americans today truly are and what they believe in. Again, we see the visibility logics of both internally focused pride and community building and externally focused descriptive norm-setting working side by side. The testimony of Brandon and William thus suggests how young conservatives are employing citizen marketing practices that bear a striking similarity to the coming out model of the gay and lesbian movement. As such examples drawn far from the field of conventional identity politics suggest, virtually any constituency that perceives itself to be an outsider group in a particular locality may see political value in increasing public visibility by having members participate in symbolic self-labeling practices.

Whether such value is truly to be gained, however, remains an open question, and the precise persuasion effects of participatory visibility campaigns have yet to be measured in any scientific capacity. Moreover, the political effects implied in these efforts are diffuse, long-term cultural shifts—that is, what Ethan Zuckerman calls a "cultural theory of change" as opposed to the more traditional "legislative theory of change"[49]—that may perennially resist quantification. However, the firsthand accounts above offer a compelling set of logics to suggest how making identities visible might indeed support the long-term interests of myriad political identity groups, in terms of both connecting members together who were previously invisible to one another and challenging external perceptions about their presence, size, and scope in the broader world.

Challenging Visible Identity Campaigns and Articulating Difference

Although the practice of mediated self-labeling has been thoroughly embraced in a wide range of political contexts—both in identity-focused social movements and beyond—it has also been the subject of criticism. In particular, the controversy over visibility politics within the gay and lesbian movement compels us to consider what may be lost when identities are flattened into a limited set of symbolic markers, as well as how critical voices might respond by symbolically articulating their own difference and dissent. By returning once more to the case study of the red equal sign profile picture campaign, we can better illuminate the innate risks of participatory mediated visibility, as well as the role of grassroots digital media practices in potentially redressing them.

Why exactly has visibility in the media become such a lightning rod of debate in the gay and lesbian community? As noted previously, the simple answer is that any visual representation of an identity group is inherently reductive, leaving intragroup diversity hidden from view. For many gay and lesbian authors and activists, a frequent concern is that the images that reach the broader public are those that are the most comfortable and nonthreatening to mainstream audiences (i.e., white, middle-class, gender normative, and scrubbed of sexuality), whereas images that are less immediately palatable (i.e., nonwhite, lower-class, gender nonnormative, and so-called sexually deviant) are left out of the equation. Along these lines, Suzanna Walters argues in her critique of visibility in television and film that "the complexity and diversity of the gay and lesbian community needs to be represented, not promoted as simply heterosexuality with a twist."[50] In the context of gay-oriented commercial advertising, Amy Gluckman and Betsy Reed observe that "the real contours of the multicultural, class-stratified gay populations are languishing in the closet, while images of white, upper-class lesbians and gay men become increasingly conspicuous."[51] Furthermore, the transgender community has often been completely left out of media visibility projects that purport to be under the LGBT banner. Katherine Sender points out, for example, that commercial marketers typically "ignore transgender people within the universe of the gay market."[52]

Such concerns about the limitations of assimilationist forms of media visibility mirror the ongoing debate within gay and lesbian activism regarding its political priorities and overall direction. In particular, the issue of same-sex marriage, for years one of the chief focal points of mainstream gay activism organizations such as the HRC, Equality across America, and the National Gay and Lesbian Task Force, has been criticized for privileging the priorities of wealthier (and disproportionately white) members of the community who stand to gain the most from legal marriage rights. Alexandra Chasin, for instance, contrasts the rights-based discourse of mainstream gay and lesbian identity politics with the social justice discourse of queer and feminist radicalism, arguing that the former fails to secure the well-being of the full range of sexual minorities.[53] Sarah Warner, also a radical queer critic, decries the recent political focus on gay marriage as a worrisome example of "homoliberalism" that primarily benefits the existing social and political status quo while leaving aside many who exist at the margins of society.[54]

Considering these critiques, it is no surprise that the red equal sign—produced by the HRC organization as a way of symbolizing support for marriage equality rights—was not universally embraced by all LGBT citizens as a representation of collective identity. Rather, as the campaign took off on Facebook and other social media sites in 2013, critical voices began to interject that the HRC's political agenda did not reflect that of the entire community, and therefore its logo failed to serve as an accurate visual label of who they are as political subjects. Writing for the *Huffington Post*, the blogger and activist Derrick Clifton makes a typical critique, noting fears that many "economically well-off, able-bodied, gender conforming, non-immigrant and white (read: relatively privileged) gay and lesbian Americans will disengage from the many other institutional and social changes necessary for full inclusion of LGBT communities . . . just the sight of the HRC logo recalls that scary possibility."[55] Clifton goes on to point out that the HRC has had a poor track record in terms of working on behalf of transgender people, further fueling the controversy regarding the organization's focus on gay and lesbian marriage rights. Thus, for some, the HRC red equal sign represented a narrow and partial image of the community, not unlike the assimilationist media representations that have been chastised by visibility critics.

And how did the red equal sign campaign's critics respond? One obvious route was to simply not participate; in the context of the campaign's enormous popularity on sites like Facebook, this in itself could be seen as a conspicuous act of dissent. Another option was to voice one's objections directly in the public discourse, as Clifton had done with his article. However, a surprising twist of the red equal sign campaign was that some of its critics tackled the image head on, posting modified versions to articulate their concerns in the representational space of the profile picture campaign itself. This intriguing turn of events signals how participatory media platforms are sometimes used to challenge—as well as to support—collective visibility campaigns that seek to unite identity groups under a shared repertoire of symbolic labels and markers.

Indeed, assuming that such connective actions simply advance a single political agenda in a wholly linear fashion ignores the complexity of peer-to-peer digital culture. Instead, we must always be attentive to the ways in which citizens seize digital phenomena as spaces for their own message making, whether by creating entirely new modifications or, more commonly, by selectively forwarding ones that they find online. In some cases, this activity may directly counteract the communicative intentions of the initial content producers. As Jenkins points out, the standard model of viral promotion "places an emphasis on the replication of the original idea, which fails to consider the everyday reality of communication—that ideas get transformed, repurposed, or distorted as they pass from hand to hand."[56] In this case, the HRC image did not remain static, but was instead modified and transformed in countless ways that were both supportive and critical. We can thus think of those who participated in the spread of the red equal sign meme on Facebook as grassroots intermediaries of the HRC's particular brand of LGBT movement identity: whereas some acted to strengthen the brand by wholeheartedly standing behind its logo as a marker of personal identity, others worked to challenge it through interventionist strategies of digital remix and parody. Referring to the latter as "critical digital intertextuality," Richard L. Edwards and Chuck Tryon note how such practices stem from the logic of textual poaching, which figures like Jenkins have used to describe how active and agentic audiences repurpose media content for their own cultural self-expression. As Edwards and Tryon emphasize, "critical digital intertextuality requires a remix mentality—it is no longer about just decoding or opposing the preexisting content of a visual entertainment, it is about contesting it, using bricolage techniques to challenge its status as a transparent media text."[57]

Over the course of my interviews, I spoke with several respondents who posted customized versions of the red equal sign that were intended to contest the HRC's message as well as its role in broader LGBT politics, effectively reworking the campaign as a space of reflexive critical discourse. For example, Elizabeth, who is transgender, explained that although she wanted to participate in a profile picture campaign to support LGBT rights, "I wasn't super keen on adopting an HRC symbol. . . . It seems like they sometimes sell out trans people, and that they're not always at forefront of activism. I've heard it called 'corporate gay.'" With this critique in mind, she decided to change her profile picture to an image she found online of a red equal sign joined with a transgender symbol (which combines symbols for male and female). She went on to explain that her goal in posting this compound image was to increase visibility for transgender citizens, something that she felt was not being achieved in the HRC's public advocacy and marketing efforts.

To express their own qualms with the red equal campaign, two other respondents described posting parody images of the red equal sign as a kind of subtle visual protest. Sean, who described being a member of radical queer organizations, used the mathematical "greater than" symbol with a red background as "a sort of sarcastic rejoinder to the equal sign . . . in that the equal sign has kind of been a really, I would say a facile shortcut for equality." As he explained, his goal in posting this parody image was to encourage critical thought about the HRC and its politics: "Even though they claim to be working for equality, they tend to ignore portions of the greater LGBT community, and tend to work towards more politically palatable goals. . . . I was mostly just trying to get people to think about what it was that they were doing." In a similar fashion, Riley, a transgender man, changed his profile picture to a modified parody image that depicted the red equal sign superimposed with a popular online meme of a cartoon seal shouting "GAAAAAAYYYY." As he explained, he did so because "I felt very annoyed about the whole situation. I was being very sarcastic by doing that." Although he described posting the image as "just a little bit of satire, just trying to be like 'I think this is sort of silly,' " he connected it to a serious critique of the HRC and its marriage rights focus. Riley further echoed Elizabeth's concerns about the diminished status of transgender citizens within the HRC's political efforts: "We are also a part of the LGBT community, and we're not being helped in almost any way by this unfortunately. . . . I have an issue with the fact that this particular campaign, this particular issue, doesn't address a lot of other social justice issues that need just as much work as this one does." Thus, through the use of humor and parody, Riley sought to poke holes in the campaign's unified image of visibility and spread the word that not every member of the LGBT community identifies with the HRC logo and what it stands for.

As these accounts underline, peer-to-peer networked digital media have an innate malleability[58] that enables alternative voices to emerge in participatory media campaigns and challenge their meanings as well as the politics of the initial creators. Although such critical interventions may only exist on the periphery of efforts like the red equal sign campaign, they are important to consider in the context of broader debates regarding the democratization of power in participatory culture. As Bennett argues, "the empowerment offered by distributed, networked digital communication . . . warrant[s] an important adjustment to media hegemony theories."[59] In the case of the red equal sign campaign, although we can recognize the hegemonic force of the HRC in working to unify the LGBT and ally communities behind its own mainstream version of assimilationist equality, we can also appreciate how the organization's use of digital networked media introduced significant possibilities for counterhegemonic discourse.

Interestingly, it is this very malleability inherent in digital content such as memes that the HRC credited with the red equal sign campaign's broader success. According to Anastasia Khoo, "we certainly could have taken a much different approach to try and control the campaign or to brand it more tightly, but success relied on allowing people to make our logo their own and feel like they were part of something bigger."[60] As noted at the beginning of this chapter, this resulted in myriad manifestations of the red equal sign meme that were supportive in character, often involving the juxtaposition of the logo with favorite popular culture icons. This sort of personalized connective action, Bennett and Segerberg argue, has the potential to support the advocacy agendas of political organizations while broadening citizen participation.[61] However, Natalie Fenton and Veronica Barassi are more pessimistic about the value of this activity for political organizations, positing that "self-centered media production practices, which are promoted by social media, represent a challenge to the construction and dissemination of political messages that are born out of the efforts and negotiations of a collective" and may therefore "represent a threat for political groups rather than an opportunity."[62]

The case study of the red equal sign campaign demonstrates both dynamics simultaneously. Although the memeification of the image may have catalyzed popular participation in a way that was amenable to the organization's goals (as Khoo implies in her laudatory assessment), more critical variations like the ones posted by Elizabeth, Sean, and Riley were expressly intended to undercut the HRC's message. However, considering that organizations like the HRC have come under fire for claiming to represent an entire identity group while sometimes leaving certain portions underserved, opportunities for members of these groups to go "off-message" in their circulation practices may be democratically valuable. Indeed, such personalized communication via participatory platforms appears to open much-needed spaces to interrogate the role that large and powerful issue advocacy organizations like the HRC are playing in broader struggles related to identity politics and beyond.

As more political constituencies turn to mediated self-labeling tactics as a means of promoting their interests, these tensions are important to keep in mind. By their very nature, participatory visibility campaigns seem to require a stock set of representative symbols (i.e. images such as the equal sign, slogans such as "It Gets Better," etc.) that enable the coordinated articulation of identity in the public realm of appearance. However, as we have seen in this chapter, the mediated representation of collective identity is a practice that is continually fraught with tension, as particular subsets come to be more conspicuous than others and intragroup diversity is hidden and suppressed. Although this may be a perennial problem for visibility-centered approaches to citizen marketing, the critical interventions and parodies deployed by some participants in the red equal sign campaign suggest one possible avenue of redress. As critics like Walters insist, true visibility must be wholly inclusive for a group to potentially improve the conditions for all of its constituents. Thus, the more that members of identity groups can make visible the full range of their diversity through customized and personalized forms of self-labeling, the more these groups can potentially "enter the scene of history" in their entirety and make their multifaceted presence and political interests known to the broader society.

Conclusion: Refining Visibility Tactics in the Digital Age?

The recent multiplication of digital technologies that allow for message customization thus promises to make mediated visibility more truly participatory—and inclusive—in the future. However, it would be a mistake to assume that such a process will be easy or automatic or that the technology itself will somehow "solve" the problem of diverse representation on its own. On closer inspection, in fact, the case study of the HRC's red equal sign and its critical rejoinders alerts us to how certain attributes of networked digital media may challenge the capacity of social actors to meaningfully expand the field of representation and avoid the trappings of a standardized and reductionist identity politics.

First, although the meme culture fostered by networked digital platforms includes a healthy amount of purposeful variation and even dialogical critique, its simultaneous emphasis on the speedy replication of trending content seems to push back against this sense of endless possibility in favor of an altogether more conformist mode of circulation. In the red equal sign campaign, the critical parodies and remixes such as those discussed above were, in reality, dwarfed by the overwhelming number of more agreeable iterations, effectively drowning out these countermessages in the public discourse. This imbalance is readily observable in the broader press coverage of the campaign: of the dozens of mainstream news articles written about the red equal sign meme, almost none mentioned any critical intervention coming from members of the LGBT community.[63] Overall, the critical variations of the campaign from transgender citizens and other critics appeared to have relatively little impact on the subsequent discourse, in contrast to the massive amount of public attention received by the many versions of the meme that more or less fell in line with the HRC's message.

To be clear, pointing to such an imbalance is not the same as claiming—incorrectly—that the spread of Internet memes is somehow a mechanical process devoid of human agency. As Shifman reminds us, "the depiction of people as active agents is essential for understanding Internet memes, particularly when meaning is dramatically altered in the course of memetic diffusion."[64] However, although such alterations in meaning can and do occur with great frequency, their relative impact is not always necessarily equivalent, even as peer-to-peer Internet technology itself significantly levels the communication playing field. The reason for this would appear to be more cultural than technological. For instance, there is a palpable dynamic of voguish trend-following in contemporary meme culture that celebrates the ability of Internet users to become part of whatever is popular online at any given moment and perennially stay "in the loop." In the context of political memes, this would suggest that once a digital artifact has gone viral in the culture by reaching a certain critical mass of widespread exposure and emulation, its spread may take on more and more of a bandwagon-following character while pushing more critical or alternative iterations to the margins. Indeed, we have seen how the self-labeling visibility tactics detailed in this chapter often follow the logic of a bandwagon effect, setting norms for others to follow and imitate in seemingly contagious fashion. The race to join in on the next big online trend, so it would seem, leaves little time to engage in the critical reflection necessary to meaningfully alter its course.

Thus, we can appreciate how the two primary attributes of Internet memes as a cultural practice—variation and replication[65]—exist in permanent tension with one another when it comes to the peer-to-peer circulation of politically charged content. Whereas the former promises the articulation of a near-infinite heterogeneity of viewpoints and experiences, the latter may swing the pendulum back toward the direction of ideological uniformity. Under this set of cultural conditions, meaningfully expanding the representation of political identities beyond already-established or standardized forms of visibility is by no means impossible, yet such efforts may often face an uphill battle.

In addition to the threat of a limiting conformity, a further dynamic of networked digital media that poses challenges to these sorts of efforts is its strong emphasis on concision. Although notable exceptions exist, much of the content that goes viral in the contemporary social media landscape is remarkably succinct, squeezed into the confining spaces of hashtags and 140-character tweets or quick-bit videos and images designed to grab the momentary attention of users scanning an endless stream of updates. Again, this is a matter of social and cultural rather than innately technological factors, as companies such as Twitter make conscious decisions to limit the size of the content that circulates on their networks as a way of catering to users who seek constant stimulation and novelty in easily digestible bites. In political contexts, this dynamic can be observed in everything from the sloganeering of hashtag-based actions to the logo-centered profile picture changing campaigns discussed previously. To be clear, the attribute of concision in mediated political discourse is not new to the social media age—as addressed in previous chapters, the field of political marketing has long relied on packaging politics with simplified symbolic appeals (slogans, logos, sleek images, etc.) that critics have chastised for trivializing complex issues. However, the digital context appears to heighten this dynamic of reduction and simplification to a considerable degree, as political content is funneled into short bursts of data that are made to circulate rapidly from peer to peer.[66]

In discussions of political identity and visibility, this emphasis on concision in representation raises clear concerns. As noted above, many critical scholars have pointed to the framework of identity politics more broadly as dangerously oversimplifying for the way in which it privileges a single characteristic over others and obscures the complexity of multidimensional oppression. As feminist scholars of color such as Kimberlé Williams Crenshaw have argued, this complexity demands an intersectional understanding of issues like race, gender, sexuality, class, and ability—that is, one that recognizes how various forms of oppression often work in concert to differentially impact social subjects, and one that does not force individuals to choose which aspect of their identity is the most politically germane.[67] By contrast, the mediated identity labels considered in this chapter would appear to push in the opposite direction, pinning citizens down to a limited set of immediately legible symbolic markers that ultimately fail to represent their full complexity as human beings.

In the red equal sign campaign, the HRC logo effectively lumped together all LGBT citizens and their allies into a single visual brand that erased many important differences, much in the way that critics like Williams Crenshaw describe. Even critical interventions, such as posting the greater than symbol in place of the equal sign, faced the dilemma of how to complicate images of visibility with the often-reductive language of Internet memes, opting in this particular case for pithy satire but perhaps losing some of the complexity of its intended meaning along the way (indeed, in our interview, Sean had to explain the political critique imbedded in the greater than image because I had difficulty understanding it on my own). As this example suggests, the emphasis on quick and easy legibility in networked digital media may challenge the ability of citizens to articulate nuanced critiques of identity politics and other issues that defy simplification.

Does it follow, then, that complex concepts like intersectionality cannot be expressed in terms of profile pictures, hashtags, and other digital bits? Do the reductive dynamics of online meme culture always translate into reductive forms of visibility that ultimately privilege some members of a group over others in ways that replicate existing inequalities? Although such questions have no easy answers, there is cause for optimism. As I argue in more detail in Chapter 6, connecting concise symbolic packets to complex political realities is a difficult process, yet its potential success hinges on the interpretive and hermeneutic work of individual social actors. In other words, forging political meanings out of Internet memes and other packaged accoutrements of citizen marketing is not simple or straightforward, but requires a considerable amount of reflexive thought and praxis. There is no reason to rule out the possibility that, say, a potent symbol of intersectionality or subaltern coalition building will begin spreading on Facebook or Twitter feeds at any point in time—particularly given the inherent malleability of digital technology that places few formal limitations on peer-to-peer expressivity—although the process of retrofitting such complex ideas into the concise textual and visual languages favored by popular digital platforms would require significant creative and interpretive labor. This, then, constitutes a major challenge moving forward for citizens who wish to complicate standardized displays of visibility on social media (and elsewhere) that threaten to conceal as much as they reveal about the thorny matter of social and political identity.

In this chapter, we have seen how the coming out model of mediated self-labeling has emerged as a core logic of participatory citizen marketing practice, stemming in many senses from the cultural visibility projects of the gay and lesbian movement and expanding into myriad efforts to elevate the public presence of groups that have been hidden, obscured, or underestimated by common-sense perceptions of social and political reality. By entering the public realm of appearance, as Arendt puts it, these groups gain significant opportunities to both compel the attention and sympathy of the broader public and foster feelings of communal pride and belongingness that are crucial for long-term mobilization. At the same time, the simplified, branded labels of identity that are typically used to achieve such ends risk suppressing the complex intersectionality of contemporary political struggle (along lines of race, gender, sexuality, class, ability, and more). Furthermore, the trendiness and succinctness that often characterize the culture of memes and social media seemingly make the nuanced refinement of visibility tactics all the more difficult, although a range of digital practices such as parody and remix hold the promise of expanding the representation of identity beyond the most limiting constrictions.

In the following chapter, we will see how the promising yet fraught practice of conspicuous public self-labeling has also emerged as a key element of participatory promotion in the context of electioneering. Rather than coming out from the cloak of invisibility, however, these practices utilize displays of group membership and identification to perform a fanlike enthusiasm for candidates and parties in ways that encourage others to follow suit. At the same time, the reductive nature of these expressive displays of political allegiance raises its own set of concerns, not only in terms of trivializing complex issues but also in terms of dividing citizens into polarized, cordoned-off "teams" for branded partisan promotion.

4

Political Fans and Cheerleaders

Promoting Candidates, One Brand Evangelist at a Time

During the 2015 general election in the United Kingdom, one of the most high-profile moments of candidate promotion came not from the parliamentary parties themselves, but from a 17-year-old student named Abby Tomlinson, who started a trending hashtag on Twitter declaring her admiration for the Labour leader Ed Miliband. #Milifandom was quickly adopted by thousands of like-minded, mostly young British voters, who then continued Miliband's memeification by posting his portrait as their profile pictures and circulating photoshopped images of him fashioned as popular culture icons like Superman. As Tomlinson commented to the press about the hashtag campaign, "We just want to change opinions so people don't just see the media's usual distorted portrayal of him, and actually see him for who he is. Ed is just a great guy and how many other politicians have a fandom? Zero."[1] This focus on attempting to influence peers by circulating promotional political messages places the #Milifandom phenomenon squarely within the citizen marketer approach. Yet despite Tomlinson's claims of ingenuity, this kind of grassroots outpouring of candidate fandom has become an increasingly common fixture on the contemporary political landscape. For instance, across the Atlantic, youthful fandom on social media and elsewhere was a major narrative in the landmark 2008 presidential campaign of Barack Obama, who inspired a strikingly similar episode in which a college student produced a fawning amateur viral video called "I Got a Crush on Obama" that racked up millions of views on YouTube.[2] Such exuberant brand evangelism may vary widely from candidate to candidate, yet the broader pattern of its emergence signals that the citizen marketer approach is becoming deeply ingrained in the institutional political systems of advanced Western democracies.

Indeed, #Milifandom was only the opening salvo in a series of social media–fueled political fandom episodes in the United Kingdom. Not to be overshadowed by the Milifans, supporters of the Conservative leader David Cameron quickly responded with the hashtag #Cameronettes in an attempt to foster a rival fan base. Like his Labour-boosting counterpart, the university student who launched the #Cameronettes hashtag explicitly identified peer persuasion as his motivation, stating that "the Left tend to be extremely active on social media and hence can influence the younger demographic. . . . However, it's nice that a pro-Conservative trend has emerged. Hopefully it can highlight the positive record that the Conservatives can be proud of."[3] After Cameron's party won the election and Miliband left his post as Labour leader, an even more high-profile social media effort emerged in support of Jeremy Corbyn as Miliband's replacement. Although Corbyn was widely perceived to be an underdog because of his outsider position in the Labour party, he eventually won the leadership election while riding a wave of social media enthusiasm. This included the prominent hashtag #JezWeCan, which at its height was being tweeted 25 times a second, as well as popular photoshopped memes depicting Corbyn as heroic figures ranging from James Bond to David Beckham to Obi-Wan Kenobi.[4]

In the United States, Corbyn-mania found a parallel in the grassroots social media efforts to boost the 2016 presidential chances of Bernie Sanders, a similarly insurgent candidate who also faced an uphill (and ultimately unsuccessful, yet remarkably close) battle to capture the top position in his party. The hashtag #FeelTheBern, initially popularized by a small group of grassroots activists who were independent of the formal Sanders campaign, became an astounding viral success during the Democratic primary election cycle. In the summer of 2015, it outpaced the official campaign hashtag of Sanders's opponent Hillary Clinton by a ratio of more than two to one on Twitter.[5] A few months later, another Sanders supporter created a promotional effort called #BabiesForBernie, which involved the sharing of user-generated photos of babies dressed by their parents in the candidate's signature white hair and glasses. After the initial photo inspired a legion of imitators on Instagram, #BabiesForBernie was spun off as a Facebook group, as well as a website that sold T-shirts and onesies featuring the slogan—all with the goal of helping Sanders fans spread the meme and the message to their peers.[6] Like many of the above examples, #BabiesForBernie surfaced online without the involvement of any formal campaign or party organization. Rather, effusive supporters took it upon themselves to produce—and, more crucially, to circulate and selectively forward—content that essentially functions as political marketing for their favored candidates. In such cases, promotional labor emerges from "unofficial" political spaces of popular culture, or what Richard Iton calls the vernacular tradition of political discourse.[7] Moreover, the dissemination of both "official" and unofficial hashtags supporting candidates like Sanders and Corbyn was largely peer to peer, powered by citizens working on their own accord as grassroots intermediaries for their preferred candidate brands.

It was at a 2008 Obama rally that I first personally witnessed the spectacular intensity that this kind of political fandom can reach in the contemporary electoral realm. The setting was downtown Philadelphia, days before the much-anticipated Pennsylvania Democratic primary election. The scene surrounding Obama's campaign speech in Independence Park felt more like an arena rock concert than a traditional campaign event: tens of thousands of Obama's followers filled the grounds adorned in myriad T-shirts and other accoutrements declaring their love and veneration for the candidate. Some featured the typical logos and slogans of the campaign's formal marketing outreach, yet many others were distinctly vernacular popular culture artifacts celebrating Obama as a kind of folk hero. Reverent portraits abounded, including the famous blue-and-red "HOPE" image created by the street artist Shepard Fairey, along with phrases like "Barack Star" and "Obama is my Homeboy." However, what struck me most about this dizzying visual cornucopia of pro-Obama sentiment was not its presence at the rally itself, but its potential endurance when the gathering dispersed and these citizens returned to their lives with the exuberant slogans and portraits still splashed across their bodies. I thought too of the deluge of photos and status updates that would inevitably be posted to social media sites like Facebook as a commemoration—and amplification—of the rally's dazzling visual displays. To a great extent, scenes like these are the modern equivalent of the citizen-powered parades of the 1840 William Henry Harrison campaign, only this time, the popular zeal for candidates need not be wholly directed by party leaders and campaign operatives (even as these institutional actors reap the promotional benefits). Furthermore, these spectacles of fandom and promotion are not limited to traditional political spaces like formalized rallies and gatherings, but rather spill into the fine meshes of everyday life—Twitter and Instagram feeds, as well as the vistas of city streets.

The citizens who disseminate these kinds of candidate-promoting messages via memes, hashtags, T-shirts, and more represent an updated version of what Kathleen Hall Jamieson calls the "surrogate message carrier," the very kind that had supposedly disappeared as 19th-century participatory campaign spectacle gave way to the top-down, professionalized, mass-mediated political marketing of the 20th century.[8] Yet as we saw in Chapter 2, citizen marketing practices have been intensifying rather than waning in recent times because of important cultural as well as technological shifts. As political expression and cultural self-expression have become increasingly connected, showing one's support for a candidate can effectively double as a statement about one's own deeply held sense of identity. However, although this dynamic of cultural identification helps account for why citizen marketer efforts like #Milifandom and #BabiesForBernie often emerge from the vernacular spaces of grassroots popular culture, it is important to remember that they do not exist in a vacuum.

Rather, as I will argue, the culture of contemporary political fandom is inextricably linked to broader strategies of candidate image building that have been developed over many decades within the professionalized field of political communication. The intermingling of institutional politics with popular culture fandom and celebrity worship is thus as much a top-down phenomenon as it is a vernacular response on the part of ordinary people. To appreciate this point, one need look no further than Donald Trump, who successfully leveraged his fan base as celebrity host of the hit reality TV show *The Apprentice* into a political movement that jumpstarted his ascendance to the U.S. presidency in 2016. Indeed, today's political fans and cheerleaders are fully embedded in a hybrid media system[9] that combines new media logics of user-generated content, memes, and virality with older mass media logics that facilitate candidate packaging and image craft.

The idea of political fandom and the expressive culture that surrounds it forms a central theme of this chapter, which explores the relationship between citizen marketing practices and the field of electoral politics. It is a notion that is epitomized not only by the youthful social media users who gush about their favorite candidates by creating celebratory Twitter hashtags and YouTube videos—as well as knocking their opponents with parody, derision, and attacks—but also by the many thousands who share this content with peers in the hopes of making it go viral to impact the election process. Undoubtedly, the incorporation of citizen marketing practices into the realm of institutional politics has important consequences for those who plan and manage such election campaigns. Yet the class of political communication professionals who seek to shape citizen marketing activity for their strategic gain is only a partial focus of this chapter. My primary concern, rather, is with the citizens themselves who engage in these practices as a means of asserting their own political will. As we will explore, these practices are changing what it means to become a part of the democratic process and are opening new avenues for popular involvement through the symbolic performance of enthusiasm and affective identification. At the same time, their boldly partisan character may also exacerbate tensions in everyday social interaction and contribute to the tribal-like cultural divisions that increasingly characterize contemporary civic life.

However, before we examine what motivates citizens to outwardly express their fandom for electoral candidates and parties to their peers—as well as the implications of this activity for social and civic life more broadly—we must first consider the larger context from which contemporary political fandom emerges. Unlike the identity-based social movements discussed in the previous chapter that have traditionally congealed through an interplay of bottom-up and top-down processes, the modern electoral system has long depended on candidates and parties actively and vigorously promoting themselves to voters via elaborate marketing appeals. Thus, to contextualize the phenomenon of citizen marketing in the electoral realm, we must examine how it has developed in relation to professional electoral marketing campaigns, which are often some of the most expensive and heavily strategized in all of the promotional industries.

Electoral Political Marketing in the Digital Age

At a time in which the mantra *your consumer is your marketer* sits at the conceptual center of the promotional industries as a whole,[10] the process of professional electioneering is undergoing significant transformations. One area of development that has been well explored in recent scholarship is the formal extension of institutional political marketing into networked digital spaces and the adoption of electronic word-of-mouth strategies. Kreiss traces this shift back to as early as 2000, when U.S. presidential campaign organizations "increasingly used the Internet to fashion supporters into the conduits of strategic communications" and worked to "take advantage of existing social networks to create a new 'digital two-step flow' of political communication."[11] In her historical review of Internet-based electioneering in the United States, Stromer-Galley shows how campaigners have enlisted supporters to amplify the reach of their carefully crafted campaign messages, often by building official interactive applications on platforms such as Facebook and YouTube.[12] Although she notes that Internet campaigners may struggle to manage and regulate the two-step flow of persuasive messages on digital platforms that afford users a large degree of individual agency, she argues that they are becoming increasingly adept at using digital communication technologies to "direct and control citizen-supporters to work in concert to achieve campaign goals."[13]

In a study of U.S. digital political consultants, Serazio likewise emphasizes how campaigns are using social media as a venue for "seeding campaign messages . . . operatives have obviously long recognized the authenticity and persuasive power of word-of-mouth, but the new media tools enable strategists to harness it toward electoral ends—leveraging social networks and maximizing shareable content."[14] Referring to this phenomenon as a key element of what she calls "citizen-initiated campaigning," Rachel K. Gibson observes in the UK context that parties often focus on creating "tools for supporters to cooperate in distribution of the party message."[15] As these accounts underline, contemporary political campaigners are not content to sit back and hope that their candidate goes viral through grassroots fan hashtags and memes, but instead actively work to shape the flow of electronic word of mouth by providing tools and templates to transform supporters into disciplined brand evangelists.

A further reason why political campaign organizations are so eager to launch participatory marketing efforts online is the value of the personal data they generate. Unlike any earlier communication technology, every action and interaction on the Internet produces data points that can be collected, stored, aggregated, analyzed, and packaged for a variety of strategic purposes. As early as 2000, U.S. electoral campaign organizations began using Internet outreach to gather personal information from voters to segment them into smaller groups and send them tailored messages.[16] As Kreiss and Howard note within the context of the 2008 U.S. election, campaigns have used techniques such as building Facebook applications to "help make social networks both visible in the form of data and productive for fundraising, mobilization, and voter identification efforts."[17] Presumably, all present and future digital outreach efforts launched by electoral campaign organizations will be culled for the valuable information they reveal about the citizens who engage in them. However, it would be a mistake to assume that professional electioneers are only interested in fostering peer-to-peer digital communication to mine data for their top-down tailored marketing efforts. Although this may be an important "back-end" goal contributing to the growth of the phenomenon at an institutional level, the peer-to-peer circulation of campaign promotion also holds the seductive promise of paying its own viral dividends.

The tactical use of campaign supporters as both data-mining opportunities and message conduits raises important questions about the extent to which professional electioneers are controlling and managing voters via digital technologies. Stromer-Galley argues that by fashioning what she calls "controlled interactivity" through campaign-administered websites and social media apps, political institutions undercut opportunities for genuine democratic participation and instead exploit citizens as mere "pawns on a chessboard" to win electoral battles.[18] In other words, rather than use interactive technologies to lend supporters a voice in the formation of policy proposals, campaigns opt for a transactional approach that positions them in a purely service role. Kreiss takes a more mixed position in this debate, arguing that "theorists who see a dystopic form of elite management and network optimists who see enlightened collaboration as the consequence of changes in technologies miss the hybridity of a form of organizing politics that combines both management and empowerment."[19] Moreover, Kreiss emphasizes that critiques of elite control neglect to consider the fact that "the interests of campaigns and the publics they mobilize are *aligned* most of the time" (italics in original) and that "many supporters not only accept but embrace" this transactional service role, "given the basic goal alignment between these campaigns and their supporters: the objective is to defeat rivals, not remake democracy."[20] As Kreiss suggests, this alignment of goals helps account for why many Obama supporters in 2008 used online expressive tools that were outside the control of the campaign, as well as those that it had designed itself, to lend support to their favored candidate.

This point about citizens feeling so closely identified with a politician that they are willing to be used instrumentally as message conduits and brand evangelists begs a crucial question: how did they get that way? If campaign supporters see themselves as members of a candidate's team, ready to "fight the vital 'ground' and 'air' wars," as Gibson puts it,[21] how are such teams formed in the first place? Part of the answer appears to lie in the strategies of digital campaign operatives, who often focus on the community-building dimensions of their official online applications. For instance, in an ethnography of U.S. congressional campaign offices, Jessica Baldwin-Philippi finds that "campaigns are increasingly requesting that supporters circulate campaign material to their own networked publics" as a means of "facilitating interpersonal connections" and strengthening attachments with communities of fellow supporters.[22]

However, to fully interrogate this question of candidate team building, I argue that we must look beyond the specific domain of formalized digital campaigning and consider the broader dynamics of fanlike cultural identification that have increasingly characterized the political landscape of advanced Western democracies such as Britain and the United States. To a significant extent, this phenomenon can be traced back to the work of professional political marketers, who have over time developed sophisticated strategies for branding candidates as objects of celebrity worship through a wide range of media appeals. However, the scholarship on political fandom (and political marketing more broadly) has tended to emphasize a more reciprocal relationship between politicians and the public, with emotional attachments to candidates arising from the interpretive activity of citizens rather than from a wholly top-down packaging process.[23] As we turn now to the issue of how political fandom is mutually co-constructed, the logic of peer persuasion often adopted by citizen marketers in electoral contexts begins to come into sharper focus: rather than acting as mere conduits for campaign messages, those who outwardly symbolize their support for favored candidates work to model authentic emotional attachments in ways that they hope will grow the enthusiastic fan bases of their revered team leaders and, in turn, their electoral prospects.

Celebrity Candidates, "Cool Politics," and Political Fan Cultures

As discussed in previous chapters, the field of institutional political marketing has a long history of using mass media appeals to sell appealing images of candidates to voters. In the 1950s, a series of slick television advertising spots may have been enough to package candidates for mass consumption. However, by the 1990s, candidate image making had grown to include myriad forms of popular cultural engagement that blur the boundaries between political communication and celebrity media culture. Street uses the term "cool politics" to describe how public officials have come to perform the role of popular culture celebrities as a means of igniting the interest of today's voters—particularly young people who have largely become cynical about political institutions and who have moved away from traditional party affiliations. For Street, "this is not just a matter of being popular, but being popular in a particular way. [Candidates] want to be stylish in the way that stars of popular culture are stylishly cool."[24] In the 1997 British general election, for instance, the Labour leader Tony Blair made a point of associating himself with popular rock stars such as Noel Gallagher of Oasis in high-profile media photo-ops.[25] In the United States, Bill Clinton's 1992 presidential campaign contained numerous moments fitting this trend, most famously when he played the saxophone on the *Arsenio Hall Show* and spoke with young voters on MTV.[26] After Blair and Clinton's youthful support helped pave the way toward electoral victory, the strategy has been endlessly copied (although some have applied it more successfully than others).

In addition to this move toward celebrity-like popular culture engagement through television and entertainment formats, politicians have also begun to use social media to reach voters in more culturally situated contexts. As Jason Gainous and Kevin M. Wagner argue, platforms such as Twitter strengthen the ability of politicians to control their message and influence their supporters, largely because these platforms "circumvent the media's gatekeeping function" and provide a direct conduit to followers who are already likely to be receptive.[27] This suggests that the cultivation of political fandom may be becoming more effective in the social media age, as candidates use their online profiles and feeds to craft appealing images without the interference of journalists who are traditionally tasked with holding these images accountable. In an expanding and hybrid media environment, journalistic watchdog activity still takes place all the time in a variety of outlets, but Gainous and Wagner make the point that citizens are increasingly sidestepping this coverage as they self-select digital media content that conforms to their political views. In particular, they find that the most partisan-minded citizens tend to follow media from favored politicians and activist groups rather than from mainstream news outlets.[28] The result, Gainous and Wagner contend, is an empowered class of politicians who enjoy an increased capacity to sell themselves and their agendas to a receptive audience without the traditional media filters.

In the 2016 U.S. election cycle, this dynamic was epitomized by the Republican candidate and eventual victor Donald Trump, whose campaign as a whole represents a chillingly logical endpoint of the hybridization of electoral politics, popular culture, and media celebrity. Trump leveraged his immense public notoriety as a television star to amass a Twitter following in the millions that towered over that of his competitors and, in the words of the *New York Times*, mastered the platform "in a way no candidate for president ever has, unleashing and redefining its power as a tool of political promotion, distraction, score-settling and attack."[29] Labeling Trump "the Twitter candidate" and likening his success on the platform to Franklin D. Roosevelt's pioneering use of radio and John F. Kennedy's mastery of television, *Salon* noted that Trump "has proven himself as the first major U.S. politician to use social media in a way that truly amplifies his message beyond traditional campaigning."[30] Crucially, this message amplification hinges on the reciprocal relationship between Trump and his impassioned fans, who not only follow him on Twitter but also circulate and endorse his posts through retweeting, liking, etc.

However, whereas Gainous and Wagner suggest that politicians can harness the power of top-down digital media outreach to influence voters, other scholars emphasize the capacity of the political audience to more critically evaluate the process of political image making. For instance, identifying the above-noted cool politics trend as part of a broader "restyling of politics" characterized by a "new prominence of discursivity, symbol-making, and aesthetic design," John Corner and Dick Pels argue that these trends can result in "new forms of visual and emotional literacy" on the part of citizens.[31] In contrast to the pessimism of trivialization critiques that have continually dogged political image-making practices (discussed in Chapter 1), Corner and Pels suggest that this kind of affective and culturally situated engagement has the capacity to strengthen critical understandings of institutional politics in an era of increasing cynicism. Liesbet Van Zoonen offers a similarly optimistic assessment that argues "for the equivalence of fan practices and political practices, an equality that facilitates an exchange between the domains of entertainment and politics that is commonly thought to be impossible."[32] For Van Zoonen, popular culture fandom resembles partisan political behavior in the sense that both have come to hinge on deep emotional investments among audiences. Referencing the atmosphere of political rituals such as the Obama campaign rally described above, she notes that "when the party leader arrives, the scenes of crowds yelling and cheering are not so different from the sight of fans shouting for their favorite sports or movie star."[33] Although such affective bonds may be criticized as irrational or vulnerable to manipulation, Van Zoonen maintains that they lead to an affective intelligence[34] that drives both fans and citizens toward active roles in discussion, community building, and creativity.

In her analysis of the role of marketing in democracies, Scammell likewise contends that emotional responses to politicians can serve as a gateway for critical judgment rather than blind obedience. For instance, addressing the focus on candidates' personal lives and character qualities through their media appeals, Scammell posits that "personalization of politics invites us to engage our emotional intelligence and evaluate politicians by the normal standards of popular culture. It links distant high politics to the everyday."[35] Furthermore, Scammell describes a broader shift in political marketing toward the so-called branding model, which emphasizes a reciprocal relationship between brands and publics that takes seriously their interests, needs, and emotional responses. As she explains, "brand images, the marketing literature insists, emerge as much from the bottom up as the top down"[36] and are co-constructed in a way that depends on "the experience and perception of consumers, which in turn arises out of multiple and diverse encounters."[37] In other words, brand images are arguably as much a product of audiences as they are of the marketers themselves, and are successful only when they are genuinely responsive to citizens' interpretive processes of meaning making. Thus, what the work of scholars like Scammell and Van Zoonen suggests is that political fandom is not a wholly manufactured invention of elite image makers and social media–savvy politicians, but is rather to some extent an organic outcome of campaigns and candidates that meaningfully resonate with citizens on deep emotional levels through their public outreach. Scammell admits that such formula may perennially risk a prioritization of style over substance, as a multitude of trivialization critics warn, but she nevertheless insists that what she calls "good" campaigns will successfully appeal to both the emotional intelligence and the political rationality of voters in ways that foster authentic engagement and participation.[38]

Either way, the deeply felt emotional identification with politicians that scholars like Scammell and Van Zoonen describe appears to be a key ingredient in moving citizens toward a participatory marketing role in elections. At face value, it would seem that the purpose of branding candidates as popular culture celebrities would be to simply inspire a level of fanlike devotion that would motivate voters to choose them at the ballot box. However, as commercial marketers have come to appreciate, popular culture fandom does not merely end at the point of "sale." Rather, as researchers of fan cultures such as Jenkins have emphasized, fandom is inherently participatory in nature, galvanizing a wide range of expressive and creative practices at the grassroots level.[39] Banet-Weiser also makes this point in her more critical work on brand culture, noting that contemporary branding strategy is increasingly focused on moving consumers toward revering and identifying with brands enough to proliferate them through expressive forms of participatory cultural production.[40]

It would thus appear that fostering citizen-level participation in electoral marketing outreach is not simply a matter of designing the right tools to create and share promotional content. Rather, it is intimately tied to broader practices of image making and cultural engagement that forge emotional connections between voters and candidates. In other words, the fanlike devotion fostered by cool politics approaches—as well as the restyling of politics more broadly—constitutes a key ingredient for spurring citizen marketing practices in the electoral context. Without the impassioned zeal of a candidate's fans, a grassroots culture of media participation is unlikely to take off and achieve the promotional goal of going viral. Thus, from the campaign perspective, the objective is to develop a resonant brand image that will inspire supporters to do much of the promotional labor *on their own* at the grassroots level, rather than attempt to wholly direct this labor from the top down.

During the 2016 U.S. presidential election cycle, such a lesson was seemingly learned the hard way, as campaigns faced public backlash when trying to manufacture their own viral moments on social media platforms. For instance, the Hillary Clinton campaign suffered an instructive high-profile incident on Twitter in which a call for her youthful supporters to express themselves through emoji images was widely mocked. Tweeting from the official Clinton campaign account, the following prompt was given to followers: "How does your student loan debt make you feel? Tell us in 3 emojis or less." The ostensible goal was to kick-start a social media trend that would put Clinton's campaign into the Twitter spotlight, with supporters both creating their own emoji responses and sharing those made by their peers. However, many of Clinton's own followers expressed dissatisfaction with the tactic, which they saw as a form of pandering to the youth vote. For example, one Twitter user replied, "I love you but we don't need you to do this," and another sarcastically questioned, "is there a 'condescended to' emoji?" The overwhelmingly negative response was then picked up by journalists as a laughable social media "fail," with one reporter describing the Twitter reaction as "swift and merciless."[41]

At a time when the political communication landscape is inundated with irreverent expressions such as pop culture–referencing memes and jokey Instagram photos, such a backlash may seem puzzling on the surface. However, the incident points to the risk of fabricating vernacular expressions of political fandom from the top down, which may be perceived as inauthentic by those who might otherwise be receptive to the same kinds of expressions when coming from the bottom up. It also may be the case that the emoji tactic conflicted with the overall brand image that the Clinton campaign had been attempting to cultivate through its broader media outreach, which tended to frame her as the most serious-minded candidate in the race. The backlash could thus be interpreted as a case of voters exercising their emotional intelligence, pushing back against the top-down construction of political brand image when it fails to resonate. Indeed, the Clinton emoji episode suggests the challenges that campaigns face as they attempt to exert controlled interactivity, in Stromer-Galley's terms, in a peer-to-peer communication environment that depends on the participation of ordinary people who may or may not wish to play along with the script.

Many success stories can be cited as a counterpoint, as Stromer-Galley indeed does in her analysis of the 2008 Obama campaign's digital outreach.[42] What, then, distinguishes the viral successes from the fails? As I have argued in this section, the answer may only be partially attributable to digital campaign strategy in and of itself, such as the building of interactive applications and the generation of viral-ready content. To reap the promotional benefits of a grassroots fan culture that operates above and beyond the parameters of official electioneering and into the vernacular spaces of the popular, campaigns must foster deep levels of connection and identification with voters that come about through much broader image-construction processes. However, whether these public images are the authentic results of citizens' emotional intelligence or instead an insidious form of manipulation through stylized packaging and top-down message control is another question. As we will see, the lines between the two may not always be clear when it comes to the lived experiences of those who adopt the citizen marketer approach in electoral contexts.

As we shift now from the campaign perspective to that of citizens who willfully participate in peer-to-peer candidate promotion, the notion of authentic personal identification, or at least the appearance of it, becomes paramount. In exploring how citizen marketing participants understand the persuasive potential of their symbolic displays of candidate support, we find that the performance of connection and identification seems to matter much more than the circulation of any specific campaign message. Once we have outlined this logic in detail, we will then consider how these displays of deeply held, culturally situated identification feed into broader dynamics of polarization and partisanship. As we will see, these declarative symbolic practices pose significant challenges to civic life at the same time as they promise to help political fans and cheerleaders rally their side to electoral victory.

Contagious Enthusiasm: Rallying the Like-Minded and Building Momentum inside the Bubble

How exactly do citizens who circulate election-themed promotional content conceptualize the potential efficacy of these practices? What audiences do they intend to reach when selectively forwarding campaign-supportive (or opposition-denigrative) messages to their peers, and what precisely do they hope will happen as a result? In the following section, I consider how the public performance of fanlike enthusiasm for politicians is deployed for the goal of building and strengthening support among those who may already be sympathetic or receptive to the message. As it turns out, rallying one's own side in a cheerleading-like fashion constitutes perhaps the most crucial, if overlooked, logic of citizen participation in electoral political marketing.

Indeed, such a strategy may seem counterintuitive on the surface. When speaking with citizens who seek to boost their favored campaigns and candidates through media-spreading activities, a major concern that is often raised is the seeming futility of reaching only those who are already in agreement. As Cass Sunstein shows in his work on partisan blogospheres, citizens with strong political allegiances are increasingly separating themselves into communities of the like-minded, creating "information cocoons" and "echo chambers" that limit contact between those with divergent viewpoints.[43] Gainous and Wagner also find convincing evidence for this pattern on social media platforms, noting that "by allowing the consumer to pick their own network of communication, social media allows citizens to self-select their content in a way that avoids any disagreeable ideas or interpretations."[44] This apparent inaccessibility of one's opposing political camp was referenced by several interview respondents who doubted their ability to change minds with their own promotional symbolic actions. Trevor, for instance, explained that "my followers on Twitter are predominantly progressives, liberals . . . I don't really expect that we're making converts in tweeting these things because by and large the people who are following are people of like mind." Laura similarly noted that being politically persuasive on Facebook is difficult because "by now some of the people with whom I have deep philosophical disagreements have unfriended me. . . . The population of people on my Facebook page is shrinking to people who share my point of view. There aren't many of the others left."

This sense of the political opposition being out of reach is especially pertinent in the context of social media platforms, where users are given the ability to manually control their contacts and screen out anyone perceived to be troublesome or undesirable with a few clicks of a button. Researchers in Israel, for instance, find empirical evidence of politically motivated unfriending on Facebook, as 16% of Jewish Israeli users (of more than 1,000 surveyed) reported that they had severed their connection with someone because they disagreed with what that person had posted regarding the 2014 Gaza conflict.[45] Another large-scale survey study finds that exposure to divergent or "cross-cutting" political views is notably limited in online spaces that are primarily political in nature, lending further evidence to fears of a fragmented and polarized digital public (however, this was less the case in online forums that are more culturally oriented and less explicitly political in nature).[46] Furthermore, even those who would avoid self-segregating behaviors on the Internet may nonetheless be pushed into so-called filter bubbles because of the algorithmic processes of sites like Facebook. For instance, the EdgeRank algorithm filters posts from Facebook connections according to how often, and in what ways, a user interacts with them.[47] As a result, a Facebook user's feed may automatically remove or filter down posts from friends that show weaker ties in the network—such as those who have fewer political interests in common—giving an algorithmic boost to the echo chamber phenomenon that is also fueled by voluntary unfriending (in response to growing complaints about the polarizing effects of news feed filtering, Facebook released a high-profile internal study in 2015 purporting to show that "compared to algorithmic ranking, individuals' choices about what to consume had a stronger effect limiting exposure to cross-cutting content,"[48] although the study was subsequently criticized for its questionable methodology and self-serving framing of the data).[49]

Such political self-segregation also extends well beyond the Internet echo chambers and filter bubbles. When I spoke with college-age Obama supporters who wore campaign-themed shirts around their heavily Democratic college campuses (located in the urban northeastern United States), they often pointed to the political homogeneity of their immediate surroundings as a reason why they failed to see their displays as serving a persuasive function. Cassandra, for instance, recounted that "we were kind of in a bubble because a lot of people on this campus were Democrats," and Eric noted that "the group of people I'm going to be hanging out with is probably similar-minded, so wearing a T-shirt is not going to cause as much debate or discussion. . . . It's preaching to the choir." For those who circulate promotional political messages to their peers, the struggle to move beyond "preaching to the choir" may thus seem unsurmountable in both online and offline contexts.

If the people who are most likely to come into contact with peer-to-peer persuasive messages about elections are already in agreement, then what, if anything, is the point of sending them? Some might conclude that it is in fact pointless, and the echo chamber of social media is often cited by observers as a reason to dismiss grassroots expressions of candidate support as merely frivolous sideshow.[50] However, when faced with the apparent dilemma of only being able to reach the politically like-minded, some respondents whom I spoke with suggested a compelling logic: success on the electoral playing field is not only a matter of securing popular support as such, but is also tied to the relative *intensity* of this support. Therefore, it would follow that reinforcing commitment among members of a political faction would potentially strengthen its ability to achieve its collective goals, making preaching to the choir a worthwhile endeavor.

In democratic systems like that of the United States and Britain that do not require mandatory voting and whose elections are often decided by the turnout levels of one partisan base over another, this logic does seem to make sense. For instance, back in 2010, when President Obama's approval ratings were down and his party was facing the threat of losing the upcoming congressional elections to the Republicans (which indeed it eventually did), some Obama supporters spoke of trying to reinvigorate their like-minded peers through spirit-lifting promotional appeals. Rachel explained that wearing her Obama T-shirt in the run-up to these 2010 midterm elections was meant "for people who already like Obama, but have not yet considered voting this year . . . hopefully it will hype up some excitement for this election," and Ryan remarked that wearing his Obama T-shirts during this period was "directed to people who voted for him and are now scared . . . like an energy campaign. We should be wearing our shirts, posting articles on Facebook, just maintaining something." Rachel and Ryan thus framed their public displays of enthusiasm for the president as serving a voter turnout function, prodding fellow supporters of Obama's party to make the effort to continue registering their support at the ballot box.

This notion of helping to build an "energy campaign" aimed at an audience of sympathetic peers, rather than an out-of-reach opposition, was also articulated by Trevor in the context of Twitter: "By and large the people who are following me are people of like mind . . . I think the main thing you accomplish is interaction with others, and it sort of builds an energy . . . I think it's more energizing people, keeping the momentum going, and keeping people engaged." Although Trevor did not provide examples of what this "energizing" activity could achieve in a concrete sense, one could imagine what might result: in addition to potentially increasing voter turnout, such a boost in enthusiasm could translate into more campaign donations, more organizational volunteering, and perhaps more citizen marketing activity as well. However, Trevor's and others' use of abstract terms like "momentum," "energy," and "hype" suggests a diffuse and intangible dynamic at play that is not necessarily reducible to any particular measurable result. Rather, such language seems to gesture toward affective relationships between citizens and their favored candidates, forged in part through cool politics and other culturally oriented forms of appeal. In other words, the spreading of energy and momentum among members of one's own faction can be understood as an attempt to extend and multiply the emotional intensity of political fandom, strengthening a form of identification with candidates and parties that is deeply cultural in nature. Here, the promotional labor of the citizen marketer becomes the performance of a contagious enthusiasm—the cheerleader drumming up team spirit by modeling an exuberance for others to follow.

This conceptual framework of modeling the affect of fandom accounts for how the peer-to-peer circulation of campaign-promoting content can potentially serve a useful reinforcement function for an audience of the like-minded. Furthermore, it is also helpful for understanding citizen marketing practices aimed at winning the electoral support of undecided voters. In interviews with citizens who wore candidate-branded T-shirts in public during the 2008 U.S. presidential election cycle, a consistent theme was their desire to visually depict the popularity of their preferred candidate in such a way as to compel others to join the proverbial bandwagon. As the logic goes, those who witness these images of grassroots support would then be more likely to vote for the candidate because they sense broad social approval among their peers. Anthony, the president of his school's College Republicans, explained this line of reasoning in detail when accounting for why he and his fellow club members wore John McCain shirts to local campaign events in 2008:

> We want to make sure that people know the candidate has support in that area, and that's an important way to actually increase support for them. It's attractive to have a whole group of students going around wearing a T-shirt . . . people think to themselves "oh, there's energy and youth support for the candidate." Generally, it does lead to at least people considering that candidate and his or her views.

Rachel, the president of her school's College Democrats, articulated a similar logic when describing how she saw wearing Obama T-shirts as helping her candidate's electoral chances that same year:

> There's definitely this "part of a crowd" kind of mentality. When everybody's wearing the same thing, there's definitely this collective force. If they see that their candidate is popular, then they'll think "oh, well I should probably vote." They're not going to go out and waste their time to vote for somebody who's down. . . . So seeing everybody wearing something that supports the candidate kind of says "alright, yeah, this person has some support."

As Anthony and Rachel's comments suggest, this logic of persuasion is an extension of the visible identities model outlined in the previous chapter. By making their favored candidate's supporters into a more publicly visible and legible group through practices of self-labeling, they saw themselves as rearticulating political reality and reorienting social norms in a way that can subtly influence the behavior of others. In this case, however, visibility is less about stepping out from the shadows and making one's interests known to the world than about depicting the relative intensity and enthusiasm of rival political fan bases. In other words, self-labeling is used here to facilitate a bandwagon effect that could persuade undecided voters to line up behind the candidate who has seemingly captured the adoration of "the people."

This kind of momentum-building strategy is particularly ripe for critiques of political trivialization. The persuasive message being sent in this context does not appear to have anything to do with a candidate's actual policy platform.

Rather, it communicates only his or her relative popularity and level of social salience—the so-called energy surrounding the campaign. A model of encouraging voters to pick their leaders based on how beloved and cool they are perceived to be among their peers may sound to many like the ultimate example of aestheticized politics as antidemocratic mass manipulation—the very kind that the Frankfurt School theorist Walter Benjamin famously warned about in the context of the Swastika-emblazoned popular spectacles of Nazi Germany, which, in his words, provided the masses "not their right, but instead a chance to express themselves."[51] Furthermore, there have been numerous cases of candidates allegedly counterfeiting such displays of grassroots identification and enthusiasm via deceptive tactics. For instance, it was reported in 2015 that Donald Trump hired professional actors to wear Trump-branded T-shirts at a rally announcing the start of his presidential campaign.[52] In the previous election cycle, the Republican presidential candidate Newt Gingrich was accused by a former campaign staffer of purchasing roughly 80% of his million-plus Twitter followers through fake dummy accounts and follow agencies.[53] However, even if displays of grassroots candidate support are genuine rather than counterfeit, by spurring their fellow citizens to hop on political bandwagons and express their desire to be part of a victorious crowd, are those who engage in this kind of citizen marketing activity nonetheless complicit in manipulative practices?

Amanda, one of the college students I spoke to who wore Obama T-shirts during the 2008 election, expressed some sympathy with this line of critique. Although she publicly displayed her support for the candidate at the time, she admitted that "now looking back, I realize I was pretty uninformed, and kind of easily persuaded by a good campaign . . . I think I can admit now that I was naive and that I didn't know really what I was campaigning for." As Amanda explained, she got caught up in the Obama bandwagon that had surrounded her and her peer group: "I just think that everyone around me was voting for Obama, and he seemed to be this positive, young, inspiring guy . . . just seeing the frequency of it, how many people liked it, and the fact that all of my friends liked it." She further noted that "pop culture was very pro-Obama, and he had all these celebrities endorsing him and things like that . . . I know that had an effect on me." In other words, the recommendations of admired popular culture figures—one of the essential elements of cool politics, as noted previously—worked alongside the recommendations of trusted peers in persuading her to both support the candidate and participate in the promotion of his campaign. In hindsight, she regretted her focus on style over substance: "I didn't really look at the nuances of the campaign and I didn't look at the issues as much. . . . I just think that I wasn't really critically thinking."

To some extent, Amanda's candid testimony lends credence to the fears of trivialization and manipulation that haunt the political marketing spectacle in its contemporary, participatory phase. The packaging and branding of candidates as objects of celebrity devotion rather than architects of substantive policy, the focus on emotional bonds over complex issue-based appeals—all are seemingly extended and intensified by peer-to-peer forms of candidate promotion that model political support as a kind of viral trend. Amanda's story could thus be seen as pushing back against the optimism of theorists such as Van Zoonen who emphasize the critical acumen of fan cultures in both entertainment and political contexts. By her own account, Amanda did not feel that she demonstrated much in the way of emotional literacy or affective intelligence when she opted to go along with the fashionable Obama trend and leave critical thinking aside.

However, such a story is far from universal, and it would be problematic to assume that the affective intelligence of political fans—defined as "how emotion and reason interact to produce a thoughtful and informed citizenry"[54]—is merely a wishful fantasy. Rather, Amanda's account prods us to consider the potential risks, as well as the presumed benefits, of blurring the boundaries between political communication and the ecstatic brand evangelism of fan cultures. And what, in fact, are these benefits? As scholars like Van Zoonen, Street, and Scammell would be quick to remind us, the imbrication of politics and popular culture offers new pathways for political participation in an era in which formal political institutions are at risk of losing broad public involvement because of cynicism and disillusionment. For Van Zoonen, the key to overcoming civic disengagement is to determine "how politics can borrow from the elements of popular culture that produce . . . intense audience investments, so that citizenship becomes entertaining."[55] In a similar vein, Street argues that stylized packaging makes politics more accessible to citizens by reducing the difficulty of paying attention to issues and providing "cheap information" that makes engagement more likely from a rational choice perspective.[56] From humorous Internet memes and T-shirts to flashy YouTube viral videos, the expressive tools of citizen marketing do appear to help make electoral politics entertaining and accessible, particularly for young people who have grown up in a postmodern world dominated by media and popular culture. To be certain, showing support for candidates with symbolic artifacts is more fun and easy than the tedious legwork of traditional campaign volunteering (although, as I will address in more depth in Chapter 6, the relationship between these two forms of activity is cause for much controversy and debate). Although risks of trivialization and manipulation may always be lurking around the corner, it is important to recognize the role of political fans and cheerleaders in spreading the emotional investments that make the duties of citizenship an altogether more entertaining and appealing enterprise.

However, the public expression of celebrity-like adoration, cultivated by cool style and slick candidate packaging, does not fully account for the range of citizen marketing practices in the context of elections. It may go a long way toward explaining the popular spread of reverential profile-picture portraits, gushing YouTube video clips, and impassioned Twitter hashtags, but what about the other side of the political marketing coin—attacking the opposition? In the following sections, I explore the relationship between the deep, culturally grounded investments of political fans and the partisan antagonism that colors much of contemporary political discourse.

Partisan Rancor in a Polarized Public

During the 2012 U.S. presidential election cycle, one of the most memorable citizen marketing moments of the vernacular variety took place during the televised debates between Obama and his Republican opponent Mitt Romney; while the two candidates went back and forth, online partisans spun out instantaneous memes mocking the statements of their preferred candidate's rival. During the first debate, Romney's remark that he would "fire" the *Sesame Street* character Big Bird (in reference to his plan to cut funding for U.S. public television) inspired Obama supporters to memeify this perceived misstep with everything from humorous image macro memes to parody Big Bird Twitter accounts. As the digital ridicule spread across the Internet, news articles with titles like "How the 'Fire Big Bird' Meme Could Hurt Mitt Romney"[57] began to proliferate in headlines (notably, the official Obama campaign organization seized this moment and quickly produced its own Big Bird–themed anti-Romney television ads that followed the lead of the grassroots memes,[58] an intriguing example of how vernacular citizen marketing practices are now often driving official political communication strategy). During the second debate, Obama supporters circulated another round of online anti-Romney memes that mocked the candidate for using the phrase "binders full of women" when discussing his experience promoting gender equity in the workplace,[59] an incident that was echoed four years later when feminist supporters of Hillary Clinton transformed Donald Trump's "nasty woman" remark into a popular anti-Trump hashtag.[60] It was as though the ranks of these candidates' teams were ready to pounce at every turn, flooding the web with derisive jibes in the service of turning public opinion against their competition. Of course, these symbolic skirmishes cut across party and ideolological lines: for instance, anti-Obama memes had been inundating the U.S. conservative blogosphere since the candidate first became a public figure, depicting him as everything from a Communist dictator to the *Batman* villain the Joker.[61]

In the 2016 election cycle, this trend of candidate bashing at the grassroots level reached a fever pitch as Clinton and Trump, two public figures with the highest recorded unfavorability ratings in U.S. presidential campaign history,[62] became their respective party's nominee. In response to Trump's rising poll numbers, for instance, critics unleashed a torrent of anti-Trump memes and videos on social media, including "Darth Trump," a parody clip depicting him as the *Star Wars* villain, along with image macros and GIFs portraying Trump as real-world villains such as Hitler and Mussolini. Writing for the *New Yorker*, Ian Crouch notes that on social media, "Trump's critics seem game to match Trump's own hysterical hyperbole—to fight fire with fire. . . . perhaps if enough people share these memes, the thinking goes, it will act like a social-media version of Sinclair Lewis's 'It Can't Happen Here,' and wake others up to the threat of creeping fascism."[63] Not to be outdone, Trump fans proved adept at circulating anti-Clinton memes across the Internet in the run-up to the election, with the particularly castigatory rhetoric of posts from the Reddit sub-group r/the_donald (such as a meme juxtaposing a Star of David symbol with Clinton in seemingly anti-Semitic fashion, which was later retweeted by Trump himself) becoming a source of widespread scrutiny and controversy.[64] As these examples lucidly suggest, citizen marketing practices in electoral campaigns are not limited to the effusive promotion of favored candidates, but include a hefty dose of vitriol for the opposition.

This may be especially true in the context of the United States, which has witnessed a notable rise in partisan acrimony in recent years. A widely reported study by Shanto Iyengar and Sean Westwood finds that "hostile feelings for the opposing party are ingrained or automatic in voters' minds," leading to discriminatory behaviors that even exceed discrimination based on race.[65] The authors label this phenomenon "affective polarization," since it has little to do with substantive differences in political ideology and more about feelings of in-group and out-group identification. Similarly, Lilliana Mason attributes growing polarization not to divides over issues, but to an increase in cultural identification with the opposing camps of liberal and conservative. As she explains, Americans "have become more closely identified with their parties, and these stronger identities have caused them to behave in more partisan ways . . . it is not that we are angry because we disagree so strongly about important issues; instead, we are angry, at least partially, because of team spirit."[66]

Mason's invocation of sports metaphors here is apropos. The polarization of the U.S. polity into "red" and "blue" sides speaks to inherently cultural divisions that bring to mind local football or baseball rivalries. If rival political candidates represent leaders of opposing teams on an electoral playing field, as the metaphor goes, the voting public becomes the fans in the stands that cheer for their own side and jeer the other. Indeed, in the interviews gathered for this project, I heard these kinds of sports metaphors come up time and time again. For instance, when speaking with Melissa about wearing an Obama-branded T-shirt to a spontaneous rally on the night of the candidate's 2008 election victory, she compared the experience to a similar victory rally for her favorite football team: "It was like a 'we won' kind of thing. It was celebratory, sort of like when I had my Terrible Towel with me when the Steelers won the Super Bowl. It's sort of like the same thing." In a similar fashion, Kyle, a young conservative, described how he had gotten into the habit of wearing his Republican-themed T-shirts the day after his party experienced a political victory: "I'd watch the news or commentary and say 'oh, we had this little tiny victory.' At the time it sounded like a good idea to wear it. I'm also a big sports fan, so I think the best analogy is that it's like if one of my teams wins I'll wear that shirt the next day to say 'hey, we won.'" The appearance of the phrase "we won" in both Melissa's and Kyle's accounts serves as an apt illustration of the centrality of cultural in-group identification in the contemporary political landscape.

Another common metaphor that has been used to describe contemporary political polarization is that of tribalism. Mason, for instance, writes that the nature of political conflict in the United States is shifting "from one of reasoned disagreement to one closer to ethnic discord,"[67] and the political pundit and former U.S. Cabinet member Robert Reich argues that "the two tribes are pulling America apart, often putting tribal goals over the national interest."[68] The underlying critique here is that such tribal-like (or teamlike) divisions are antithetical to the ideals of deliberative democracy. As Mason puts it, "one unfortunate consequence of this identity-based polarization is that it cannot be resolved by reasonable debate."[69] Although such concerns may be specific to the American context, they have global implications as well, especially when considering the fact that U.S. political strategy and marketing practices are currently being exported around the world through the widespread use of consultants.[70]

Indeed, one can draw a clear line between the trend of identity-based polarization and the strategies of professional political marketers, from the restyling of candidates as popular culture brands for lifestyle expression to the increasing use of negative campaign ads that paint the opposing side as the reviled outgroup enemy.[71] What happens, then, when more and more citizens are brought into the fold of these marketing tactics through peer-to-peer media practices? By participating in the circulation of partisan media content—cheering on the leaders of one's team and sniping at leaders of the other side—do the citizens who engage in these activities actively contribute to the tribal-like cleavages that threaten democratic discourse? To gain a better understanding of how these tensions play out in everyday contexts, we now turn to the accounts of those who have experienced them firsthand when confronting members of the political opposition with partisan media content.

Creating Converts or Creating Conflict?

Of all of the potential target audiences for the grassroots circulation of persuasive campaign messages, the group composed of those who support the opposing side may ultimately be the most desirable. After all, winning over converts would not only increase the ranks of one's own team, but also weaken the competition. However, considering the fact that citizens' political allegiances and voting preferences are often shaped by deeply ingrained, culturally bound forms of group identification, this audience presents the most obvious challenges for persuasion as well as the greatest potential for producing conflict and discord. Furthermore, this audience may simply be out of reach for many because of the aforementioned growth of information cocoons and echo chambers[72] within a sharply polarized electorate. Nevertheless, some practitioners of citizen marketing appear to be up for the challenge.

For instance, Kenneth, who described himself as a recent convert from Republican to Democrat, spoke of "trying to get some of my friends to convert" with his posts on Twitter about the 2012 presidential election. As he figured, his large and heterogeneous online social network made him better positioned than others to break through to other side: "When I post something, a hundred people on Twitter are going to see it . . . so I've got a pretty good reach on that. So I know if I can get a message out, if it changes one or two people's opinions over the long haul, I feel like I've done a pretty good thing." Theresa offered a somewhat different rationale for her perceived persuasive efficacy via social media, surmising that although she alone did not have access to a large and receptive audience of potential converts, her position within larger network dynamics ultimately made her circulation activities worthwhile:

> Most of my audience [on Twitter] probably would agree with me . . .
> I know I have some conservatives that follow me on Twitter, but even if
> they're not going to be willing to pass it on, some of the people who are
> more liberal like me might be willing to pass it on. And then they have
> some people, conservatives . . . and the people that follow them too . . .
> If you just keep getting this out there and other people grab it and then
> they pass it on, it's just going to spread outwards to even people who
> aren't politically of my persuasion.

In other words, by forwarding it to her peers, Theresa felt that she could increase the chances that the media content she favors would eventually expand to networks of oppositional audiences that are outside of her own immediate social network.

Although digital platforms such as Facebook and Twitter may provide limited opportunities to advance beyond the echo chambers and forward persuasive messages to those with opposing views, the scene of physically embodied public space allows for more direct forms of confrontation. For instance, many young conservative respondents described wearing political T-shirts on largely liberal college campuses as a way of deliberately challenging their Democratic-leaning peers. William, who owned a shirt printed with Obama's portrait along with the slogan "NOPE," explained that he specifically wore it around the "people who were so rabid" about the candidate after the 2008 election to compel them to question their loyalty in subsequent election cycles: "the 'NOPE' [shirt] was kind of me being like 'hey, now you know that it's not that simple.' Nothing really had changed . . . people were expecting Obama is going to save the world right off the bat. It was one of those things where like it's like 'nope, it didn't happen. Time to think it over again.'" Robert wore a similarly confrontational T-shirt featuring the slogan "I Told You So" (a commentary on the perceived poor performance of the Obama administration) around Democrats on his campus for two distinct reasons: "Wearing these shirts, one, it might make them say 'hmm, maybe I should rethink things,' or two, it might make them angry. I'm fine with both."

A similar dynamic of both angering the opposition and potentially influencing them was expressed by Brandon, who, as discussed in Chapter 3, made a habit of going to Obama rallies wearing conservative shirts with Republican and Tea Party slogans. He explained, "when I go to Obama stuff, the reason is to provoke and get them fired up. . . . I like conflict. I just get a rush by going to those types of events." Brandon recounted that he had been physically threatened on several occasions at these events because of what he was wearing and that he relished this negative reaction. Although Brandon's actions appear to be largely centered on receiving a thrill from antagonizing those with whom he disagrees, he also identified a persuasive goal: "Every time I go to an event like that, I think that if one person talked to me, if one person's viewpoint was changed, I did my objective."

As such testimony suggests, interpersonal conflict is an inevitable byproduct of citizen marketing practices aimed at the political opposition, particularly (if by no means exclusively) when the forwarded content itself is hostile in nature. For some, like Brandon and Robert, this conflict can be a source of great excitement, even pleasure. However, for others, it can be disheartening and can create unwelcome and contentious interactions. For instance, Nicole, who wore Republican T-shirts around her heavily Democratic campus, noted that "some people will make a snide comment. Usually it's more just a look, or they'll roll their eyes or something like that." Crystal, another Obama supporter, described being "picked on" by a conservative peer for wearing a candidate-branded shirt and lamented that "people really try to stir something out of you."

In fact, the risk of unwelcome conflict was so distressing for some that they chose to pull back from this kind of "in-your-face" political expression in their everyday lives. Dustin, for instance, described how he came to avoid wearing his Obama T-shirt around supporters of the opposing candidate during the 2008 election cycle: "My roommate is a Republican supporting McCain, and we'd get into some pretty heated debates at times . . . so I guess that was one of the reasons why I just didn't necessarily want to always flaunt [the T-shirt] around. I just felt like it could have provoked another debate that I just didn't feel like getting into." Dustin went on explain that he would limit wearing his political T-shirt in the future so as to not impede the development of social relationships: "I think I would remain reserved, not wearing a campaign T-shirt every day . . . because I do feel like politics are extremely polarizing. And especially for people that you don't know very well, that can have a bad effect . . . that gives them a reason right off the bat to take issue with you. And I would much rather be neutral going into a relationship." In other words, he sought to resist the tribal-like divisions fostered by heated political partisanship and maintain his ability to socialize with a wide range of people. Accounts such as Dustin's thus underscore the notion that participating in electoral marketing at the grassroots level poses real threats to the quality of everyday social interaction, potentially exacerbating the tensions of a deeply fractured public.

This concern with the divisive potential of citizen marketing practices may be particularly heightened in the context of face-to-face encounters, since the display of politically charged material objects like slogan T-shirts brazenly disrupts casual social spaces that are not formally set aside for political expression and debate. When passersby cannot help but be confronted with the sight of political messages that they disagree with, they may understandably feel accosted and react with hostility. However, other scenes of interaction, such as digital platforms like Twitter, are also not immune from the discomforting aspects of partisan confrontation. Erin, for instance, described her dissatisfaction with the contentious responses she received from her political tweets:

> I used to get into Twitter arguments with members of the Tea Party, and they could get really frustrating. And it would last hours, you know, back-and-forth, like a hundred forty characters, you'd end up like tweeting and tweeting and tweeting and it would get really frustrating. So sort of attacking other views on Twitter, or promoting other views that are in opposition to someone else . . . [Twitter] is not the most effective vehicle for that.

Erin thus came to doubt the efficacy of her efforts to win over those who disagree with her on Twitter because they seemed to produce nothing but endless arguments.

One solution to this sort of unwanted conflict may be, in Dustin's words, to be more "reserved" when encountering others. Studies led by researcher Kjerstin Thorson provide empirical evidence for this pattern among U.S. Facebook users, finding that they will often suppress their political opinions on the platform because they fear how such posts will be received by friends who are not in agreement,[73] and that they often perceive political expressions on the platform to be unwelcome "rants" that merely cause discord.[74] This phenomenon can be traced more broadly to what Nina Eliasoph refers to as "political etiquette," which is particularly prominent in the U.S. cultural context.[75] Conducting her ethnographic research prior to the growth of social media, Eliasoph observes that Americans generally avoid expressing their political views in public venues because they perceive political disagreement to be incompatible with the desired order of social life. Further, she warns that "in trying to get along, and make the world seem to make sense, we sometimes develop an etiquette for talking about political problems that makes it harder for us to solve them."[76]

By contrast, the sense of personal control offered by what Papacharissi dubs the private sphere of Internet communication[77] would seemingly embolden many citizens to become more outwardly expressive of their views from behind the comfort of the screen. Yet as Thorson's research suggests, there may be limits to this as well, particularly when considering the acrimony of much online political discourse that may be similarly unwelcome. Although the citizens spotlighted in this chapter—and this book as a whole—generally go against the grain of the pattern of reticence identified by Eliasoph, it is nonetheless apparent that the bold public declaration of political opinion and allegiance may only be attractive for a limited subset of the public that is unfazed by the threat of interpersonal conflict. Indeed, there appears to be a spectrum of approaches—one that is likely related to differences in individual personality and temperament. Some, like Dustin, may be more cautious and measured in their symbolic political declarations, whereas others, like Brandon, may throw caution to the wind as they embrace a provocateur role.

In addition to holding back to maintain a sense of political etiquette, the avoidance of partisan antagonism on networked digital media may be also be achieved, as noted above, by simply hitting the "block" button. As Oliver explained with regard to his anti-Romney tweets during the 2012 election cycle, "I have a tendency to block people who are really obnoxious, because there's no sense in talking to people . . . who are so committed to Mitt Romney that they won't listen to anything negative about him." Indeed, the tools that social media platforms offer users to control the composition of their social networks can help minimize contentious and unwelcome interactions with those holding opposing political views. This trend was demonstrated in most spectacular fashion during the 2016 U.S. presidential cycle with the launch of Friendswholiketrump.com, a website that identifies Facebook friends who hit the like button for the candidate. As one reporter put it, the site "provides us an interesting and easy way to judge our Facebook friends, or create an easy 'unfriend' list."[78] Removing social media connections to block out undesired political posts may very well bring peace of mind. The cost, however, is the deepening of partisan entrenchment—the veritable sealing off of Sunstein's digital echo chambers.

One-Way Declarations Sparking Two-Way Dialogue?

It would thus appear that citizen marketing practices aimed at the political opposition inevitably result in one of two outcomes, each of which challenges democratic ideals. One the one hand, there is the threat of antagonism and rancor; on the other, there is polarized self-segregation. Surprisingly, however, some who engage in these ostensibly confrontational activities describe them as a way of initiating reasoned civil discussions about politics with people who hold divergent views. For instance, Alyssa described her Obama T-shirt as "a conversation starter. And whether some person would disagree or agree, you can get into an insightful debate and get to the core of our beliefs, and argue or agree . . . and it's stimulating." Michael mentioned a similar dynamic when describing the conversations that were sparked by his Republican candidate T-shirts: "Some people come up and usually they'll talk to you about it . . . 'oh, why do you like them?' or whatever, and you just explain what you think about their ideologies. It's just to have a conversation in general, because I like politics and hearing different people's sides of it." Here, Michael and Alyssa identified motivations for displaying promotional material for electoral candidates that move beyond the domain of peer persuasion. Although it may seem counterintuitive, they framed their one-way partisan declarations of support as catalysts for two-way dialogue.

Similarly, in the context of Twitter, Wendy explained that posting partisan content such as humorous anti-Republican videos allowed her to socialize with a diverse group of citizens whom she described as "hardcore political junkies" like herself:

> We're all really obsessed with the election cycle. Instead of picking one another because we have the same view, it's more like picking one another because we want to talk about the same issues. . . . I'm pretty open to people coming back and saying "hey, you know, whatever you posted or your comment is really stupid for the following reasons. You should look at this."

As Wendy's comments suggest, platforms like Twitter offer opportunities for communities of "political junkies" who hold a range of viewpoints to connect and engage in open-ended discussions. Although the endless arguing noted above by Erin can creep into these digital exchanges, it would be a mistake to conclude that the peer-to-peer circulation of partisan media inherently precludes the sparking of political dialogue of a more congenial nature.

Thus, it appears that the very symbolic actions that may threaten civil political discourse may also in some cases enable it. However, in contrast to the more formal deliberative processes of consensus building addressed in Chapter 1, the political talk described in these examples more closely resembles the sort of noninstrumental discourse endorsed by Mark Button and Kevin Mattson, who cite "the benefits of an open-ended version of public talk that is not constrained by means–end rationality and which enables individuals to develop the arts of political engagement and democratic citizenship."[79] Similarly, Eliasoph argues in her analysis of everyday political interaction more generally that "citizens have to talk themselves into their political ideas together, and that means having places for casual political conversation."[80] Indeed, research on casual political talk in so-called online third spaces—that is, those that allow for informal socializing and are typically themed around shared cultural interests rather than explicitly political concerns—suggests that these spaces play an important role in preparing citizens for broader political participation by enhancing understanding and a sense of civic community.[81] Although symbolically declaring one's allegiance to a candidate (or expressing condemnation of the opposition) may seemingly conflict with the spirit of engaging in open-ended exchanges for the purpose of mutual learning and connection, it appears to help at least some citizens break the ice, as it were, of political etiquette and avoidance.

In such cases, these expressive and declarative practices can be recognized as contributing to broader civic cultures, which, according to Dahlgren, form the basis of broader citizen engagement in an era characterized by the blurring of boundaries between politics and popular culture and the public and the private.[82] To be clear, this potential outcome of fostering civic agency and identity through casual political talk—in line with the culturalist framework explored in previous chapters—is different from the instrumental, persuasive goals that are typically bound up in citizen marketing practice. Depending on the individual, persuading one's peers may or may not be a primary motivation for circulating media content that takes sides in the polarized and partisan political climate, and some may be more focused on sparking open-ended conversations and learning from one another than on winning over converts. Furthermore, Todd Graham points out that the goals of online political interaction may vary according to the type of communication environment fostered by different Internet forums and venues. Graham finds that whereas spaces formally devoted to political top-ics tend to create an environment of competitive battling and one-upmanship, those that are more informal and nonpolitical are more likely to support goals of civic learning and mutual solidarity building.[83] Since the symbolic expressions treated in this chapter tend to traverse the spaces of formal political communi-cation and informal popular culture, it appears that they are pertinent for both types of processes.

As Eliasoph reminds us, it is incorrect to assume that "the purpose of activ-ism is to win battles, not to inspire general public debate and political partici-pation," and she notes that the activists that she observed in her ethnographic research "did not just want to win; they wanted to inspire broad discussion about society."[84] The testimony presented in this chapter suggests that there are citi-zens who fall on both sides of this spectrum. It is clear from the above accounts that a desire to persuade members of the other side does motivate at least some who circulate partisan media content, and the social discord that can arise from such cross-cutting encounters may be welcomed as well as feared, depending on one's relative taste for conflict. Furthermore, it is likely that the particular kinds of symbolic political actions discussed in this chapter, which tend to emphasize deep emotional identifications with one's partisan team and the vilification of the opposing side as a kind of cultural out-group, may make a transition to two-way, mutually respectful political dialogue more difficult (if certainly not impos-sible). Rather than try to pin down these practices as producing either healthy democratic interaction on the one hand or unhealthy polarization and antago-nism on the other, it is important to emphasize the agency of individual actors in pursuing a variety of outcomes. When it comes to reaching across tribal lines and targeting the political opposition, both harmony and cacophony remain possibilities.

Conclusion: The Complex Role of the Citizen Marketer in Electoral Democracy

In this chapter, we have seen how peer-to-peer citizen marketing practices have come to complement and extend the top-down promotional efforts of elec-toral campaigns, in terms of both venerating preferred candidates as objects of emotion-laden fandom and ridiculing and bashing the competition. Whether one seeks to reinforce support among members of one's side by modeling a contagious enthusiasm, capture the hearts of the undecided by fashioning per-formances of popular approval and identification, or even challenge members of the opposition to rethink their views through confrontational displays and declarations, electoral-based forms of citizen marketing follow logics endemic to the contemporary promotional industries as a whole: the targeting of specific audience segments, the harnessing of word-of-mouth endorsement for strategic purposes, etc. As we have seen, some of this citizen marketing activity may grow directly from the digital outreach efforts of the campaigns themselves, who have over time developed a variety of tools and tactics to transform supporters into conduits for promotional campaign messages in a two-step flow of influence. However, these efforts have been supplemented, even arguably overshadowed, by the grassroots promotion of candidates at the vernacular level of popular culture, a phenomenon that is not wholly independent of the broader branding and image-making strategies of campaigns, but is also largely outside of their direct control.

This kind of instrumental, manifest participation in institutional politics at the citizen level—following the terminology of Ekman and Amna[85] discussed in Chapter 1—is broadly viewed by scholars as a positive outcome for democracy, at least in contrast to the widespread political disengagement that has characterized numerous Western democracies in recent decades. As Ekman and Amna put it, the growing academic interest in identifying new forms and pathways of political participation—particularly via the Internet—is "justified by a concern about declining levels of civic engagement, low electoral turnout, eroding public confidence in the institutions of representative democracy, and other signs of public weariness, skepticism, cynicism and lack of trust in politicians and political parties."[86] However, although the impassioned candidate evangelism profiled in this chapter would appear to be a welcome antidote to such trends and even perhaps a revivification of popular democracy, it raises its own set of concerns that stem from the long-standing critiques of political marketing's controversial role in democratic societies.

As we have seen, the peer-to-peer spread of election-themed content is often fueled by a fanlike identification with candidates and campaigns that mirrors engagement with popular culture fields of entertainment, celebrity, and sports. The act of spreading online memes that depict one's favored candidate as James Bond or Superman, or the competition as Darth Vader or the Joker, has an obvious appeal for both ordinary citizens and democratic theorists. At a time when political institutions are threatened by mass apathy, particularly from the young, this kind of emotionally charged participation at the symbolic level promises to bridge the gap between political and cultural spaces and provide citizens with pleasurable ways to not only become engaged in elections, but also make instrumental contributions in the form of peer promotion and persuasion. In contrast, critics of political marketing as political trivialization would remind us that there are legitimate reasons to question whether style-heavy expressions of political fandom might crowd out space for more complex issue discussions and even manipulate citizens into embracing—and, in turn, further propagating—appealing images that can conceal as much as reveal substantive policy positions. After all, concise popular culture formats such as memes, hashtags, and T-shirts tend to leave little room for in-depth articulations of political ideas, and the frequent emphasis on candidates' personalities (and even physical appearance) in expressions of political fandom seems to follow the most reductive tendencies of elite political packaging.

In response, those who stress the affective intelligence and interpretative agency of citizens would argue that political image making is in fact a reciprocal process, and that we must therefore take seriously the authenticity, even sagacity, of expressivist responses to politicians that traverse official and unofficial spaces of political discourse. Indeed, we can identify threads of each side of this debate in the key logic of persuasion employed by the citizen marketing participants highlighted in this chapter, in which public displays of candidate enthusiasm and identification are exploited for their potential bandwagon effect. If these per-formances are authentic outcomes of citizens' affective intelligence, then they can be seen as helping to proliferate critical and moral judgements at an acces-sible level of popular culture, and thus, as Scammell puts it, connect "distant high politics to the everyday."[87] If, however, these performances are merely cyni-cal imitations of genuine enthusiasm harnessed for the purpose of artificially manufacturing "buzz"—or at worst, outright forgeries, as in the case of paying actors to wear T-shirts at rallies or using fake accounts to boost Twitter follower numbers—then they would seemingly exacerbate the trend of democratic elec-tions as manipulative and image-centric popularity contests that are emptied of substance.

In truth, it may be difficult to definitively distinguish one from the other, and both scenarios could occur simultaneously, even within a single campaign. Thus, rather than choose a side in the long-running debate about political packaging that now looms over amateur civic practices as well as elite campaign marketing, it is perhaps more judicious to articulate the trade-offs—the potential benefits as well as the potential risks—that stem from this multifaceted set of trends. In Chapter 6, I develop the metaphor of light politics in an attempt to capture this dynamic, which refers at once to both the relative insubstantiality of the politi-cal messages spread via citizen marketing practices and their buoyant ease of travel across everyday cultural spaces. As I will argue, the dual-sided concept of light politics can be extended to the citizen marketer approach as a whole, since it points to the ambivalence of expanding popular participation in politics by broadening the purview of political marketing that is notorious for its trivial-izing (and often elite-serving) tendencies. Navigating these tensions is thus a complicated affair, demanding new forms of critical literacy that can interro-gate connections between the style and substance of various citizen marketing efforts, as well as the broader power structures that surround them.

For now, it is important to reiterate that the combination of intensive, cultur-ally situated participation in electioneering at the citizen level and the attending specter of trivialization and elite manipulation is not at all a new phenomenon, at least in the U.S. context. Rather, it harkens back in many ways to the period of the mid to late 19th century, which, as Schudson notes, was characterized by massively popular banner-waving parades and rallies for parties and candi-dates, as well as a strong sense of citizen affiliation and identification with par-tisan teams. Critics would later look back on this era as a time when powerful elites corrupted the democratic system by seducing low-information voters with stylized spectacle and appeals to in-group loyalty.[88] Similar accusations could be directed toward the contemporary phenomenon of political fan cultures. However, the increased capacity for citizens as well as elites to shape the flow of campaign messages—especially, but not exclusively, via networked digital media—appears to create a far more complex and nuanced hybrid of top-down and bottom-up symbolic power.

And how, exactly, can we describe this citizen power in the institutional political sphere? As we have seen, participants in peer-to-peer candidate promo-tion are essentially positioned in a service capacity on behalf of campaigns. As Stromer-Galley points out, this arrangement tends to foreclose a participatory role in the actual formation of a candidate's policy positions and thus threatens to disempower citizens as they become mere servants of an elite political class.[89] However, the partisan supporters who volunteer their promotional labor for candidates seem eager to be used as foot soldiers in the "meme warfare" of mod-ern electioneering battles because their goals and interests are closely aligned with their campaign of choice. The potential power of citizen marketers in elec-tions, therefore, would appear to have less to do with shaping the policies of given competitors than with selecting and promoting new and alternative candi-dates who are seen to be more representative of their political identities.

It would then follow that these practices would ultimately be most impactful when pushing insurgent and outsider candidates into the public spotlight and spreading interest and energy around their campaigns. Indeed, we can glimpse this pattern in many of the prominent political fandom episodes discussed above, such as the intraparty leadership contests that were turned upside down by popular grassroots movements—on social media and beyond—to support Jeremy Corbyn and Bernie Sanders over their more establishment-oriented party rivals. In each case, the ascent of these candidates as objects of impas-sioned political fandom, as spread through Internet memes, viral videos, and hashtags, coincided with their increasing levels of voter support. Although it would be foolish to claim that such citizen marketing activity is a decisive fac-tor in determining election outcomes (indeed, in the case of Sanders, it was decidedly not), it undoubtedly makes a significant contribution to the broader promotional outreach of campaigns. Furthermore, it can help create powerful journalistic narratives of insurgent grassroots momentum that filter through the hybrid media environment.

Notably, such a framework of citizen marketing power in elections more or less adheres to the liberal marketplace model of democracy outlined in Chapter 1, in which voters buy (and, in this case, also help sell) one candidate option over another. As we have seen, this model has been frequently criticized by scholars for limiting the part played by citizens in democracy and forestalling a more collaborative and deliberative public role in the shaping of candidates' policy positions.[90] In addition, the marketplace model has also been accused of prioritizing individual self-interest over the public interest, since consumer-like citizens seemingly choose candidates who are seen as best fulfilling their own needs rather than the needs of the broader society. With regard to the first line of critique, it appears that an expanded capacity for citizens to promote candidate alternatives through participatory media interventions would go a considerable way toward making the marketplace model of democracy more democratic, even if the candidates themselves come preformed. In governmental systems dominated by a few major parties, this would appear to have more relevance for intraparty contests such as primary elections, where insurgent and outsider campaigns have more of a fighting chance, than for general elections that typically limit voter choice to a narrow set of party-approved options. In the UK context, Corbyn's surprising ascendance to the head of the Labour Party was enabled by a crucial change in the rules regarding candidate selection and recruitment; as Labour moved to a "one member, one vote" process for selecting their leadership post,[91] the door was opened for an insurgent, citizen-powered promotional effort that made the hashtag #JezWeCan into a clarion call for left-wing voters to shake up the party. If there was ever a case to be made for the power of peer-to-peer memes and hashtags to impact electoral outcomes, then the Corbyn-mania of 2015 would serve as a strong piece of evidence. Moreover, the Corbyn story underlines Anstead and Chadwick's broader point that institutional structures greatly affect the role that the Internet plays in election campaigns.[92] Depending on the structure for selecting and recruiting candidates, the power of citizen marketers to intervene in the electoral process through mediated performances of contagious enthusiasm may vary greatly from democracy to democracy.

With regard to the second line of critique of the marketplace model, it may well be the case that the trend of growing polarization and partisanship in democracies like the United States is intensified by a focus on advancing one's political self-interest when both buying and helping to sell favored candidates. As we have seen, evangelistic political fandom hinges on a deeply held sense of cultural, even tribal-like, allegiance that inherently emphasizes in-group and out-group divisions over consensus building and compromise. As citizen marketing participants take firm, declarative stands in partisan election battles and incorporate their personal candidate preferences into their self-expressions of identity, they may be pulled further into polarized camps that are increasingly hostile and closed off from one another. In this sense, we can draw a line between the marketplace model of democracy and Mouffe's framework of agonistic pluralism,[93] since both stress the centrality of contestation and factional conflict in political processes (even as they emerge from different intellectual traditions). However, one could argue that the pure self-interest of market relations and the tribal group interest of political agonism are not necessarily one and the same, with the latter being significantly more public-spirited than the former. Although the agonistic model does contrast with the consensus-building ideal of deliberative democracy, its focus on advocating for collective group interests (including those of marginalized and oppressed segments of society) also diverges from Cohen's notion of purchaser citizens[94] who are only concerned with what elected officials can do for them personally. Thus, the tribal-like agonism that characterizes political fans and cheerleaders might represent a middle position between a purely self-interested and a wholly public-interested citizenry, suggesting a compromise of sorts between the feared excesses of the politics–market nexus and the lofty ideals of deliberative democracy that downplay the persistence of conflict and social division.

Furthermore, as discussed above, the declarative agonism of citizen marketing practice and two-way processes of political dialogue are not always mutually exclusive. Rather, impassioned symbolic expressions of political allegiance and identification create a volatile spark in casual social spaces, which can catalyze both civic dialogue and discord and retrenchment. Thus, it is up to individual citizens to determine how to strike their own balance that comports with their broader goals and sensibilities. Undoubtedly, the kinds of political fandom practices detailed in this chapter have important implications for civic identity and agency-building processes, in accordance with the culturalist model. Symbolically displaying one's identification with a campaign or candidate, as well as engaging in the casual political talk that can follow from these ice-breaking expressions, may help form the preconditions for other kinds of participation in the institutional political sphere, such as donating, volunteering, and organizing. Indeed, these expressive media-based actions may support a range of outcomes above and beyond peer persuasion, and we must take this into consideration. At the same time, however, we must take seriously the instrumental and manifest dimensions of citizens' promotional labor in electoral contexts and beyond.

However, before we turn to a conversation of citizen marketing's broader consequences, including its relationship to other forms of political participation, it is necessary to explore one additional aspect of persuasive media spreading

that has risen sharply with the growth of digital social media: the strategic circulation of journalistic information. As we will see, in contrast to the blatant
cheering and jeering of candidates and parties that underwrites formal political
marketing and campaigning, the selective forwarding of news constitutes a far
more subtle enterprise that blurs the boundaries between educating one's peers
and persuading them.

5

News Spreaders and Agenda Setters

The Promotional Labor of Raising Awareness

"Make Kony famous." This was the charge of the advocacy group Invisible Children, which launched a viral video sensation in 2012 with a half-hour web documentary on the human rights abuses of the Ugandan militant leader Joseph Kony. The peer-to-peer circulation of the video, as well as related mentions of the Kony story, were staggering: one week after the clip went online, nearly 5 million Kony-related tweets appeared on Twitter, and the video itself received well over 70 million views on YouTube and another 16 million on Vimeo.[1] Although this grassroots media-spreading activity was spurred to a large degree by celebrity involvement—including Twitter mentions from pop stars such as Taylor Swift and Justin Bieber—it also depended on the participation of millions of citizens who chose to selectively forward the content to their friends and followers. In the process, they helped realize the goal set out by Invisible Children—Kony indeed became a household name.

Shortly after its astounding success, however, the Kony 2012 movement seemed to falter. Criticism of the organization's questionable practices, including accusations of fact twisting and financial impropriety, pushed its founder toward a high-profile public breakdown.[2] Furthermore, as Kony himself continued to evade capture by international authorities and media interest in the story began to wane, Kony 2012 became a symbol for the seeming impotency of trendy social media–based political discourse and fodder for endless jokes.[3] Like many of its contemporaries, the Kony 2012 campaign was focused on raising informational awareness about an issue in the public mind, rather than on any practical steps that would be needed to address it. This alone was enough to raise the ire of many who doubt the efficacy of awareness-raising as a viable strategy for creating social and political change. What difference does it make if more people know about an issue if they are unable to do anything about it? To many, the viral phenomenon of Kony 2012 was the ultimate example of a futile slacktivism.[4]

Yet is raising awareness truly such a pointless exercise? In retrospect, the impact of injecting the Kony issue into the public conversation for the first time in the Western world appears to be significant, since it directly preceded announcements by both the United States and the European Union to launch new assistance missions in central Africa. Noelle West, the communications director for Invisible Children, later commented that despite the withering ridicule directed toward her organization, "Kony 2012 created the most opportunity and movement around this issue, more than all the eight years before it combined."[5] For the millions who participated in making Kony famous by bringing the story to the attention of their peers online, this outcome may be at least somewhat reassuring.

Several years later, the power of elevating "cause célèbres" to public attention through the peer-to-peer circulation of online news was demonstrated in far more dramatic fashion in the context of the Black Lives Matter movement, which focuses on raising awareness of cases of racist police violence as a means of pushing for social justice and policy reform. Movement hashtags like #SayHerName—popularized by cases such as that of Sandra Bland, an African American woman who died in police custody following a violent arrest by a white officer that was captured on a viral YouTube video[6]—encapsulate how Black Lives Matter activists have used the networked production of fame and publicity to thrust their issues into the public spotlight. #SayHerName was tweeted more than 135,000 times within the first week of its appearance in May 2015,[7] but it did not quickly fade into the Internet dustbin like Kony 2012. Rather, the slogan became a centerpiece of Black Lives Matter activism during U.S. presidential campaign events that pressured candidates to address the movement's issues at a policy level. For instance, the activist Tia Oso explains that "I, along with the 50 other black organizers attending Netroots Nation 2015, decided we would use the platform of the Presidential Town Hall to demand that former Democratic Governor Martin O'Malley and Senator Bernie Sanders #SayHerName and address the crisis of structural racism."[8] The strategy of mobilizing around symbolic figures of martyrdom has deep roots in political activism traditions, particularly within the African American civil rights movement that was greatly defined by transformational cases like that of Emmett Till.[9] However, the networked digital environment appears to be revivifying and intensifying this approach in new and powerful ways, as the trajectory from obscurity to widespread public notoriety becomes enfolded in the rapid-fire dynamics of hashtags, memes, and viral videos.

Like all modern political information cycles, the process of making figures like Sandra Bland famous as symbols for larger structural issues involved a range of actors in the hybrid media system, including journalists, bloggers, celebrities, advocacy groups, and politicians.[10] However, in keeping with our focus throughout this book, this chapter is chiefly concerned with the part played by ordinary citizens in the networked promotion of news for politically motivated purposes, which I argue has become a key element of the citizen marketer approach in the contemporary digital context. Using their foothold in the public sphere to extend and multiply the reach of news stories that are important to them, citizens work to reshape the flow of journalistic information in ways that they hope will advance their interests and influence political processes. Later, we will consider how these goals may be complicated by the for-profit nature of much of the contemporary digital news ecosystem, as the exploitation of viral sharing to increase clicks and ad revenue can create traps of trivializing distraction. First, however, it is necessary to examine the potential political power that is embedded in the act of publicizing news content and amplifying its public reach.

Indeed, this sort of peer-to-peer awareness-raising activity has been recognized by scholars as constituting a significant form of agenda setting, a concept that describes how the selective dissemination of journalistic information can affect what issues the public thinks about and considers important—although not necessarily what positions they will then take. In the classic agenda-setting model introduced by Maxwell McCombs and Donald Shaw,[11] elite news media gatekeepers set the agendas of the citizenry by choosing to cover certain stories over others, thus elevating their relative importance in the public eye. Since then, scholars have also identified "reverse" agenda setting as a further process by which members of the public bring attention to an issue in a way that compels the news media to increase coverage, which in turn raises its public profile.[12] In any case, the priorities established in the public mind are seen as then impacting, to some extent, the agendas of politicians and other powerbrokers who have a vested interest in responding to their constituencies. Although the process is subtle, indirect, and involves multiple steps, the act of telling the public what to *think about* (even if not directly what to *think*, in a normative sense) is widely recognized as having the potential to wield a considerable influence on public policy outcomes. In other words, agenda setting is fundamentally implicated in processes of political persuasion, and thus it is necessary to consider how participatory forms of agenda setting fit with the broader citizen marketer approach that uses the promotion of select media content as a means of advancing certain political ideas and interests.

In the era of networked digital media, citizen participation in these sorts of agenda-setting processes is notably on the rise. For instance, in *The Wealth of Networks*, Yochai Benkler documents how the public outcry on the Internet about U.S. Senate majority leader Trent Lott's racially insensitive comments in 2002 created a high-profile issue in the press where there was virtually none previously and ultimately led to the politician's resignation.[13] A research group led by Ben Sayre observes a similar pattern of activity in the gay rights movement, noting that citizens who were unsatisfied with mainstream media reporting about the issue of same-sex marriage took to social media to circulate their own "corrective" coverage as an act of issue advocacy and protest.[14] To capture this emergent phenomenon across diverse contexts, Papacharissi offers the term *networked gatekeeping* to describe "the process through which crowd-sourced practices permit non-elite and elite actors to co-create and co-curate flows of information."[15]

One aspect of networked gatekeeping that has become particularly prominent in recent years involves the crowdsourced publicizing of myriad protest movements that have swept across the world, themselves largely fueled by e-mobilization via networked digital platforms. In some cases, protest movement members and supporters focus their promotional labor on creating particular cause célèbres—a well-known example is Neda, a victim of military violence during the 2009 Iran protests who became a leading symbol for the so-called Green Revolution through the viral circulation of a YouTube video capturing her tragic final moments.[16] However, it is often the protest movements more broadly that are made famous by networked gatekeeping practices, as participants and sympathetic observers selectively forward documentation of actions like rallies, demonstrations, and occupations as a means of raising awareness for the political causes that inspire them. In a study of Twitter use during the 2011 Egyptian revolution, Papacharissi notes how protestors and their allies drew strategic attention to the uprisings in Tahrir Square by spreading on-the-scene reports to an audience around the world. As Papacharissi emphasizes, such crowdsourced information flows were driven not by norms of journalistic objectivity, but by a marked sense of political advocacy, as they freely mixed fact with emotionally charged opinion: "network gatekeeping practices documented events and permitted opinion expression in a manner driven by an overwhelming show of solidarity" for the Arab Spring revolutionaries acting to oust Hosni Mubarak as the Egyptian president.[17]

This notion of ordinary citizens contributing to the political goals of protest movements by extending the reach of their mediated documentation is also demonstrated in a large-scale digital network analysis study led by Pablo Barbera, which tracks the flow of tweets emanating from the Gezi Park protests in Istanbul in 2013 and the 2012 "United for Global Change" demonstration that involved participants of the Spanish Indignados and the Occupy movement. The researchers find that a "critical periphery" of supportive Twitter users (i.e., those outside the core group of protestors and organizers) were instrumental in spreading news of the protests to a wide online audience as well as spurring mainstream media coverage that raised further awareness of the events. As they conclude, "by expanding the audience of messages sent by the committed minority, the periphery can amplify the core voices and actions, and thus provide a way for larger numbers of online citizens to be exposed to news and information about the protest, even (or especially) in the absence of mass media coverage."[18] This peripheral, yet crucial, news-spreading role also emerged as a key theme of a smaller-scale interview study I conducted with Caroline Dadas on Twitter use in the U.S. Occupy Wall Street movement. One respondent in the study described herself as a "signal booster" when retweeting live streams from the Occupy protest sites, explaining that her goal was to expand informational awareness about the movement to peers who may have been uninformed or unsympathetic.[19] The signal booster thus serves as an apt metaphor for the promotional labor of peer-to-peer news spreading more generally, as citizens amplify awareness of events, groups, and individuals they consider important as a means of compelling both attention and consideration from broader publics.

As the growing body of research suggests, asserting a degree of control over the flow of journalistic information is becoming an increasingly widespread and significant dimension of Internet-based activism at the level of the ordinary citizen. In contrast to an earlier era in which major newspapers and broadcast networks held a near monopoly over journalistic gatekeeping, the advent of networked digital media holds the promise of democratizing the news-spreading process and providing opportunities for citizens to participate in decisions about what stories count as important. David Tewksbury and Jason Rittenberg refer to this phenomenon as "information democratization," explaining that "when citizens repost or retweet links, they are distributing news content and becoming directly involved in the exposure of others ... the Internet has created a significant space for direct channels of citizen influence on information exposure."[20] As the authors suggest, those who take on this news distribution role often represent so-called issue publics, who focus on the niche areas about which they hold strong opinions and investments. In her discussion of social media news aggregation and filtering, Papacharissi similarly stresses the politically driven nature of selective news-spreading practices, stating that citizens "inject their perspective into the news spectrum through the practice of readership and promotion of news items they deem important. . . . Since prominent presence in a news agenda grants issues credibility and visibility, social media news aggregators allow citizens direct access to an agenda they could previously only be passive observers of."[21] Here, Papacharissi gestures toward the crucial connection between social media news circulation and the viral marketing model, as participants take on a promotional, endorsement-like role when passing on news media content that they favor and wish to see spread across digital networks.

The rise of what has been called the "attention economy" would appear to elevate the relevance of this peer-to-peer news promotion to a significant degree. In the fields of business and marketing, this term has been used to describe how grabbing an audience's attention has become a more desirable currency in the context of the information surplus created by myriad digital media outlets.[22] Journalism, too, must compete in this information-rich, attention-poor environment. As Luke Goode notes, "visibility and attention, if not information, remain scares resources in the online news sphere."[23] Under these conditions, the act of drawing the attention of one's peers to certain pieces of news and information over others thus takes on an enhanced significance. By lending publicity to favored pieces of journalistic content through various selective forwarding activities (linking, sharing, etc.), citizens contribute to the production and shaping of attention—a much-valued resource in an age of information overload. Furthermore, following the reverse agenda-setting model, such activity may have significant consequences for subsequent public discourse—even holding the promise of impacting the policy-making process.

All of this does indeed suggest that the act of spreading news to one's peers is deeply intertwined with the promotion of one's political opinions and agendas—that is, the core concern of citizen marketing practice. Perhaps the most appropriate comparison for the signal booster role may be the work of public relations, an aspect of marketing communication that has long employed the selective dissemination of factual information for strategic and persuasive purposes. Since the days of the early 20th-century pioneers Ivy Lee and Edward Bernays, public relations practitioners have labored to bring public attention to particular items of news that are beneficial to their clients, often crossing the line between journalism and bald-faced promotion in the process.[24] Whereas professional public relations agents have traditionally focused on the news production stage by targeting elite gatekeepers to get favored stories in front of media audiences, citizen news spreaders largely make their intervention at the stage of distribution. However, the outcome appears similar—the creation of attention for information that one has a vested interest in pushing onto the public. In a sense, we can think of signal-boosting news spreaders as microlevel publicity agents who work on behalf of their favored political causes, organizations, and figures in the trenches of the attention economy.

Surely such a formulation is bound to make many uncomfortable. Although the field of public relations has done much to distance itself from its historical roots in propaganda (a word, as noted in Chapter 1, that was regularly used as a label for PR until its association with mass deception and totalitarian rule forever tarnished its meaning[25]), this negative reputation still endures. By contrast, the professional journalistic enterprise has become imbued over time with ideals that are strongly opposed to the notion of using the selective dissemination of information to promote one political agenda over another. Framing the citizen-level circulation of news as a public relations–like act of information "spin" may therefore be seen as a bold affront to these ideals, diminishing a sense of journalism as a disinterested tool for creating a well-informed citizenry. In the following section, we will explore the tensions that arise between the high-minded educational aspirations of journalism and the persuasive agendas of citizen marketing, and then complicate this binary by examining the increasingly partisan nature of the contemporary information environment more broadly. From there, we will turn to the voices of citizen news spreaders and signal boosters who selectively forward political information to their peers through a variety of channels, and investigate how they navigate these tensions in everyday lived contexts.

Journalism and Political Partisanship: An Uneasy, and Intensifying, Relationship

In her analysis of news curation and sharing on social media platforms, Papacharissi raises questions about the democratic potential of these practices, arguing that "the avenues for involving audiences are riddled with binary choices that echo marketing practices and are far removed from deliberative models. Granting a news item a virtual 'thumbs up or down' confines audiences to unrealistic, predetermined, and polarized positions."[26] In other words, her concern is that the selective forwarding of news requires citizens to take an all-or-nothing stand on the content in question, either lending their personal "word-of-mouth" endorsement by incorporating it into the digital expression of their own identity or rejecting it altogether. However, although the like and share buttons may represent relatively confining modes of response to news content that foreclose complex dialogical and deliberative modes of engagement, such practices are richly imbued with what we can think of as curatorial agency, a concept defined in Chapter 1. In agonistic fashion, citizens use these tools to alter news flows in ways that contest existing public agendas and push new collective concerns and priorities to the forefront. As we will see later in firsthand testimonies, decisions to amplify the reach of particular issues and narratives through social news curation often represent deliberate, well-thought-out, and sustained attempts to impact the political world.

Nonetheless, Papacharissi's point about the polarizing dynamics of peer-to-peer news circulation is crucial, alerting us to the challenges that such trends pose to normative visions of democratic civics. Tewksbury and Rittenberg likewise address the polarizing tendencies of information democratization, positing that the expanding capacity for audiences to control news flows through social media can lead simultaneously to "increased citizen involvement" as well as "decreased social cohesion."[27] In the transition from professional mass media gatekeeping to audience-powered networked gatekeeping on the Internet, citizens may only come to have knowledge about the select partisan issues that are shared among networks of like-minded peers (as noted in the previous chapter, this outcome may be further reinforced by the filter bubble phenomenon, i.e., the algorithmic construction of news feeds on sites like Facebook that tends to prioritize posts from connections with similar interests).[28] For Tewksbury and Rittenberg, the feared result is that citizens who are ensconced in polarized news circuits will become "ill-prepared to debate and act on the important issues facing a nation," even as the information environment itself becomes more diverse and inclusive of a range of priorities and perspectives.[29]

Thus, the concerns raised by scholars about citizen-directed news flows driven by partisan political investments draw our attention to how these trends come into conflict with well-entrenched ideas about the proper role of journalism and information exchange in democratic societies. In keeping with modern professional journalism's emphasis on ideological objectivity, scholars often emphasize the value of impartial knowledge building in preparing an informed citizenry to participate in democratic processes. In Dahlgren's civic cultures model, knowledge is included as one of the five outcomes of mediated political interaction that foster a sense of civic identity and engagement with public issues at a broad level; as Dahlgren puts it, "referential cognizance of the world is indispensable for the life of democracy."[30] Disinterested knowledge sharing is also deeply implicated in the normative framework of deliberative democracy, which stresses how the rational (i.e., fact-based) consideration of public issues between citizens leads to instrumental problem solving and the building of consensus.[31] Thus, Habermas centers his discourse ethics on "the cooperative search for truth," positing that reasoned argumentation and debate must subject all claims to tests of factual validity and guarantee that all participants share a baseline of objective common knowledge.[32]

However, although such ideals are widely shared in the scholarly community, some have nonetheless emphasized that the professional journalistic enterprise has continually failed to live up to the goal of disinterested knowledge sharing in practice. For instance, according to the so-called propaganda model of journalism offered by Edward S. Herman and Noam Chomsky, the systematic political biases of the Western news media—including the wholesale avoidance of particular stories that are unflattering to certain powerbrokers and governments—effectively serve an ideological function to "manufacture consent" and shape public opinion. As an archetypical example, they cite the near blackout of American media coverage of the Indonesian genocide in East Timor, which they argue is attributable to Indonesia's role as a U.S. military ally.[33] In addition, Bennett emphasizes that all journalism carries a degree of bias regardless of its pretensions and that the American norms of journalistic objectivity developed in the 19th and 20th centuries are far from universal. Rather, point-of-view-based advocacy journalism is standard practice across much of the world, and was also the dominant form of journalism in the United States in the 18th-century period of the partisan press.[34]

Such critical perspectives are buoyed by the fact that in the past few decades, the news environment has become more and more transparently ideological in nature. In the United States in particular, there has been a marked revival of the older partisan press model. As Simon Sheppard puts it, "under pressure from new communications technologies . . . the nation is returning to the traditional model of two separate sociopolitical worlds that derive their news from partisan sources."[35] For instance, the rise of opinion-heavy cable news networks such as the conservative-targeted Fox News and the liberal-targeted MSNBC has left little separation between the worlds of journalism and political advocacy. Matthew Levendusky notes that "while traditional news outlets still emphasize balance and objectivity, these partisan media outlets provide a more one-sided take on the day's events. Ordinary citizens can align their news consumption with their partisan and ideological leanings by watching these shows . . . news is no longer simply information; it can now be a reflection of one's political beliefs."[36] As Bruce Williams and Michael Delli Carpini note, the increasing emphasis on opinion and commentary in professional news is closely tied to the trends of increasing audience fragmentation and economic competition, as journalistic outlets attempt to differentiate themselves in a crowded news landscape by pushing unique point-of-view analysis to the forefront.[37]

This dynamic is even more pronounced on the Internet, where the growth of largely separate conservative and liberal blogospheres has led to the information cocoons[38] (discussed in the previous chapter) that Cass Sunstein similarly decries for polarizing the citizenry. In recent years, the partisan blogospheres have been augmented by a new generation of ideologically charged viral news sites that are primarily geared to social sharing on platforms such as Facebook. For instance, *Upworthy*, a news aggregator and publisher founded by the left-wing activist and former MoveOn executive director Eli Pariser, claims on its website that "sometimes a story opens a door to a new and better world. . . . They change our hearts, our minds and sometimes even, the world. *Upworthy* tells those stories, and helps them break through."[39] In fact, it was Pariser himself who first raised concerns about the filter bubble effect in a widely cited book,[40] and his *Upworthy* venture has been ostensibly aimed at breaking through the boundaries of polarized networks with appealing viral content that can infiltrate and persuade a range of publics (in *Upworthy*'s case, the direction of influence leans heavily toward the left-progressive end of the political spectrum). Agenda-driven, and agenda-driving, news sites like *Upworthy* are thus tailormade for spreading political influence through the peer-to-peer selective forwarding of political information, bringing the practice of advocacy journalism squarely into the age of viral media.

In some cases, these social sharing-driven news sites have an agnostic, or even downright hostile, stance toward basic accuracy and factuality when catering to their hyper-partisan audiences (who also function as their distribution networks). During the 2016 U.S. presidential election cycle, the popularity of so-called fake news sites on social media became a major cause for public concern, with *Buzzfeed* reporting that fake news articles had actually surpassed mainstream news articles in terms of Facebook engagements (i.e. shares, reactions, comments. etc.) in the run-up to Trump's victory. Sites such as the right-wing *End the Fed*, whose articles made bogus claims about rival candidate Hillary Clinton's ties to ISIS among other falsehoods, became some of the most widely-shared providers of election "news" on the Facebook platform. Although the *Buzzfeed* study found that this kind of misinformation was more prominent on conservative sites, there were also notable examples on liberal sites, demonstrating how the phenomenon crosses ideological lines.[41] The rise of this kind of fake news on the Internet fits squarely into the broader pattern of political information becoming more and more a means of advancing partisan ends; presumably, this fallacious content is preferred by its readers not only because it plays into their ideological biases, but also because it can be used to push their political agendas through sharing it, irrespective of the truth.

Beyond the growth of both real and fake partisan journalism on web and other outlets, the increasing popularity of political satire media—part of a broader trend of blending politics and popular culture and entertainment[42]— has also greatly contributed to the ideological charge of the contemporary information landscape. Research shows that formally labeled news is now only one of many places where citizens gather information about their world, and satire programs (like *The Daily Show* and *The Colbert Report*, for example) are often cited for their value in increasing the political knowledge of their audiences.[43] At the same time, however, the satire format is free from the norms of objectivity that surround traditional news reporting, and its practitioners often incorporate a hefty dose of editorial commentary into their humor. In their work on political satire television, Jonathan Gray, Jeffrey P. Jones, and Ethan Thompson point out that "satire becomes a potent means for enunciating critiques and asserting unsettling truths that audiences may need or want to hear."[44] This critical and ideological edge was perhaps demonstrated most dramatically in the case of the Egyptian satirist Bassem Youssef, whose *Daily Show*–like program was pushed off the air by Egyptian authorities in 2013 (and again one year later) after he mocked supporters of the government.[45] In Western media, some of the most popular satirists are beginning to play the role of de facto political leaders: the former *Daily Show* host Jon Stewart, for instance, staged a large-scale rally in Washington, D.C., in 2010 that was seen as a left-wing response to a similar right-wing rally led by the Fox News pundit Glenn Beck.[46]

A third trend that has contributed to the ideological thrust of the contemporary information environment is the rise of web-based citizen journalism. Although this term has many definitions, it is used most commonly to describe the amateur creation of news content such as eyewitness testimonials and on-the-scene mobile photos and videos.[47] Although citizen-level "witnessing" may be driven by a variety of factors, including simple happenstance, it has been particularly embraced by communities of political activists. As C. W. Anderson notes, this sort of open-access, participatory journalism has roots in the Indymedia movement, which formed in 1999 in the context of the antiglobalization protests in Seattle and is "characterized by a strong political agenda."[48] The Indymedia model, as Anderson terms it, frames news as "political ammunition"[49] for protest movements and other struggles and is deliberately agenda setting in nature.

As discussed previously, the wave of social media–fueled protest movements like Occupy and the Indignados have relied on citizen journalism to document and publicize their activities; furthermore, they have drawn on a supportive periphery of signal boosters to disseminate their self-reports to wider publics and amplify attention to their causes. The witnessing dynamic of citizen journalism has also become a key strategy of the Black Lives Matter movement, as bystanders of racially charged police violence are called upon to take video of these incidents and upload them to the Internet to raise awareness of the ongoing systematic crisis. As Fredrick C. Harris explains, "unlike the images of brutality that sparked outrage in the past—photographs of lynch victims hanging from trees during the age of Jim Crow or newspaper images of brutalized black bodies lying in a coroner's office—we are now able to witness and document police violence as it happens." Specifically, Harris cites videos such as that of the chokehold death of the African American Eric Garner in New York City, which was captured on a handheld mobile phone and became the flashpoint for myriad protest actions as it circulated online.[50] Such user-generated web videos, catalysts for what has been called the new American civil rights movement, demonstrate in dramatic and visceral fashion how citizen journalism is helping lead the shift toward a more transparently activist-minded and agenda-driven news and information landscape.

Whereas citizen journalism is often used to support the political projects of grassroots protest movements and traditionally marginalized groups, a fourth trend contributing to the partisan character of the political information environment comes largely from top-down flows of social media. As Gainous and Wagner demonstrate, an increasing number of citizens are choosing to consume political content directly from favored politicians and major political organizations via their online accounts, rather than from professional news outlets. This includes not only explicitly promotional material like campaign advertising messages, as addressed in the previous chapter, but also strategically chosen political news and information (indeed, the two are often interlaced). Gainous and Wagner label these "one-sided information flows" that support a single ideological viewpoint, in contrast to the "two-sided information flows" represented by objectivity-minded journalistic professionals, and emphasize the enhanced power of institutional political elites to drive informational agendas within these polarized networks.[51] At their worst, the authors warn, one-sided social media flows can circulate strategic misinformation among partisan publics without the watchdogging of fact-checking journalists, and can thus do serious damage to the cultivation of an informed citizenry that is prepared to deliberate with a shared base of accurate and verifiable knowledge.[52] In the 2016 U.S. presidential election cycle, for instance, Donald Trump spread misinformation to his supporters on Twitter with notable ease (sometimes with the help of the kinds of hyper-partisan fake news sites discussed above). At one point, Trump posted an infographic falsely claiming that 81% of homicides of U.S. whites were committed by blacks (the true statistic was only around 14%), which was then retweeted by his followers more than 5,000 times.[53] In such contexts, the citizen signal boosters who magnify the reach of one-sided information flows take on the role of conduits in two-step flows of elite influence, underscoring Stromer-Galley's argument that controlled interactivity[54] by elites constitutes a significant element of a digital landscape that is often uncritically framed as a wholly democratized and even playing field.

Thus, in contrast to the highly centralized gatekeeping of what Williams and Dell Carpini call the "old media regime" of broadcast networks and major metropolitan newspapers,[55] the digital information environment is fragmented, crowded, and inundated with the ideological priorities of politicians, pundits, bloggers, satirists, and everyday citizens. The explosion of partisan information networks and outlets in the digital age has created an environment in which the notion of an agreed-upon set of essential, factual news items of the day may be largely disappearing; in particular, the rising prominence of outright fake news on social media has led commentators to announce the "post-truth" era.[56] At a time when a wide range of partisan media advance competing sets of issue priorities to the public, and when audiences retreat into polarized information bubbles that reinforce their opinions and screen out information that would lead them to question their beliefs,[57] the notion of a "referential cognizance of the world" is by extension becoming increasingly relative. However, despite the overwhelming growth of partisanship (and even utter propaganda and deception) in journalism and political information in a variety of formats, the concerns raised by scholars about the need to build a well-rounded and informed citizenry through disinterested knowledge building are not theirs alone. Rather, for citizens who participate in the selective forwarding of political information to their peers, these tensions and anxieties are often highly palpable. In the following section, we will explore how those who engage in the practice of news spreading understand their experiences and contend with the seemingly divergent roles of informing their peers about issues and influencing their priorities and agendas. As we will see, the notion of educating one's fellow citizens is sometimes framed as a more ethical alternative to the citizen marketer framework of explicit peer persuasion, although the blurred boundaries between knowledge sharing and agenda setting form a complex site of negotiation.

Negotiating Educational and Persuasive Frameworks of Citizen News Spreading

When interviewing citizens who post politically oriented content on social media platforms like Twitter, the conversation often turned to the fine line between spreading knowledge and pushing a point of view. Whereas some respondents ardently embraced a marketing-like model of peer influence, others displayed a notable resistance to the idea of telling one's friends and followers what to think. However, following the contours of agenda-setting theory, telling them what to simply think about was often a much more comfortable position to take. For instance, Valerie explicitly distanced herself from a model of persuasion while emphasizing her role in helping to educate others through her political Twitter activities:

> Typically, when I do post about politics and stuff, it's actual factual information that I've been able to verify, not just like random opinion . . . my goal generally with things that I post is I always try to make something that will help others be more informed about the political cycle . . . and then those people can from that make their own determination. They can say "oh this is good," or like "oh no this is not" . . . I'm not just putting like random commentary out there and it's being taken as a fact.

Here, Valerie's words capture the unease that some feel about spreading political content that appears to privilege partisan spin over factual accuracy. Her notion of sharing factual political information via Twitter as a means of letting others come to their own conclusions, rather than swaying them in one direction or another, evinces how this educational framework offers a contrast to a marketing-like persuasion approach. By describing her activities as an informational public service to her friends, Valerie was able to articulate how her politically oriented tweets could meaningfully impact others while avoiding the potential distastefulness of manipulation. Following such logic, citizen news spreading would appear to serve as an ethical alternative to the citizen marketing approach, in which the determination to sway others would lead one to wallow in the disreputable murk of "random opinion."

Indeed, many respondents who adopted this kind of educational framework when discussing the act of selectively forwarding news stressed their respectful attitude toward the decision-making capacity of their peers. Donna, who tweeted often to her followers about the U.S. presidential election in 2012, remarked that her motivation was to "nudge them to research something they heard, then form their own opinion." As she explained, providing her friends with enough information to make their own decision about the election was more important to her than convincing them to back any particular candidate: "I'd love to see everyone have an informed enough opinion to feel confident in voting. Too many people don't because they just don't understand or don't want to make a bad choice. . . . I think [my tweets] could lead people to educating themselves on the candidates, which in turn could give them the confidence they might need to do the voting deed." Donna's point about her desire to help others develop into informed participants in the democratic process highlights how this educational framework aligns with Dahlgren's concept of civic cultures, as well as related models that emphasize civic agency building through informal cultural interaction.[58] Notably, these models tend to eschew a sense that such interaction may be predicated on a desire to advance certain partisan viewpoints; rather, they position peer-to-peer political discourse as fostering values of democratic citizenship at the broadest of levels. Dahlgren, for instance, posits that civic cultures "suggest the need for minimal shared commitments to the vision and procedures of democracy, which in turn entails a capacity to see beyond the immediate interests of one's own group."[59]

Although the civic cultures model may have a great deal of applicability to contexts of peer-to-peer news spreading, the line between fostering democratic citizens and political partisans is not always so easily drawn. For instance, Gabrielle, who uses Twitter to draw her followers' attention to issues that she strongly cares about, offered a revealing account of how educational and persuasive motivations can become thoroughly entangled. As she noted, "I tweet a lot about women's rights, labor stuff I'm pretty active on, GMO-labeling . . . things that aren't being talked about or shared in the mainstream media. . . . It's stuff that by sharing it, it's possibly educating people on stuff if they read it." Gabrielle then continued: "It does have a power to sway. . . . You find something like that, and it's like 'well why aren't people talking about this?' You know, 'this is something we should talk about.'" As Gabrielle's comments indicate, "educating people" and "sway"[ing] them" are not necessarily mutually exclusive goals in the minds of those who spread political information via peer-to-peer channels. Her effort to spotlight issues that she feels are not receiving enough attention elsewhere in the media is an archetypical case of reverse agenda setting. Here, the ideological thrust of this activity comes into sharp focus—creating awareness about issues favored by certain constituencies is inextricably tied to promoting their advancement on the broader political stage.

Samantha, who posts her own amateur news clips on YouTube in addition to sharing the clips of other users, elucidated this nuanced dynamic in detail. Samantha explained that she strongly identifies as a citizen journalist, and when discussing the videos that she posted about the 2012 election that edit together clips of candidates' speeches, she initially appealed to the ideal of fostering an informed citizenry in a disinterested fashion: "I really tried to keep it balanced and leave the question for them to decide, rather than lead a person to a conclusion." However, when asked about her interest in using her most recent video at the time (covering the Republican presidential candidate Mitt Romney's tax plan) as a means of spreading political influence, she commented,

> I think it definitely is me trying to persuade, because I am setting up an argument . . . I tend to do that more with issues that I think are very important. That's where the bias comes in, it's the issues that we highlight. So chances are I'm going to highlight the inequality of Mitt Romney's tax code because I think it's rooted in inequality. You're probably not going to see me dissing Obamacare on YouTube.

Thus, although Samantha asserted her respect for others' decision-making capabilities when sharing political information, she admitted that her news-spreading efforts were fueled by an underlying agenda of spreading partisan influence. Regarding her various posting and sharing activities, she at one point stated that "if it did get more votes for Obama, that would be great." Although some of Samantha's work on YouTube may be more accurately characterized as the production of advocacy journalism, her remark that "it's the issues that we highlight" succinctly captures how the act of selectively disseminating existing political information to one's peers aligns with the persuasion-oriented goals of the citizen marketer approach.

Importantly, this focus on spreading news and information seems to avoid the potential distastefulness of explicit, bald-faced promotion, while still opening a space for citizens to exert their political priorities on others. As Samantha continued, "personally I'm for Obama in a big way, but I think that you lose credibility if you just sound that way. So I think the most successful way is to present a case and leave it to the reader to decide." Here, she suggested how taking an ostensibly objective and educational tone on platforms like Twitter provides a more reputable cover for a tacit persuasion agenda, since directly telling people what to think can come across as more alienating than merely highlighting factual information for them to think about. Participatory agenda setting can thus be thought of as offering a more polite, less invasive form of citizen marketing—one that may have particular appeal for those who are wary of coming on too strong in their efforts to push their political views on their peers. Indeed, research on the social norms of political Facebook use suggests that explicit "efforts to persuade or recruit others politically" are widely perceived to be inappropriate.[60] As Eliasoph points out more generally, so-called soapboxing has long had a negative reputation in countries such as the United States that are bound by a sense of political etiquette.[61] Thus, the seemingly lighter touch of news spreading and agenda setting can come across like a more respectful and well-mannered alternative to outright soapboxing. Yet it can also be viewed as insidious—partisan promotion disguised as disinterested knowledge sharing, a kind of Trojan horse of peer political influence. Like the public relations practices that it resembles in certain key ways, this kind of strategic peer-educating activity becomes vulnerable to critiques of insincerity, even deception (whether the information itself is deceptive, as in the case of fake news, is another matter). At worst, it could be accused of that dreaded word propaganda—the historical forbearer of public relations—which has come to be defined as "coercion without the appearance of coercion."[62]

However, the intermixing of information exchange and partisan promotion via participatory media platforms is not necessarily always a matter of artful sleight of hand. Sometimes, it can be upfront and transparent. Over the course of the interviews for this project, I heard many respondents reference an educational framework when discussing their circulation of content that explicitly pushes a political point of view. For instance, some who posted red equal sign images as their Facebook profile picture as part of an advocacy campaign for same-sex marriage rights (discussed in Chapter 3) cited an interest in spreading informational awareness of the issue as a preliminary step toward enacting political change. Considering that there was no actual information included in the red equal sign image itself, employing an educational framework here may seem counterintuitive. However, Gianna, one of the participants in the campaign, stressed that her new and novel profile picture grabbed her friends' attention and opened the door for information exchange to take place. She explained that her intended audience for the red equal sign symbol was friends who "don't pay enough attention to what's happening in politics, and on the Supreme Court level," and she went on to claim that she had success in using the image as a springboard for disseminating information about the marriage equality issue: "I did have a couple people who posted and were like, 'why did you change your picture and what does it mean?' And then you get the opportunity to send them a link or to tell them that, you know, this is going on and that this is a milestone." No one could accuse Gianna of feigning objectivity in this instance; rather, raising awareness about the marriage equality issue and making a statement in support of it went comfortably hand in hand.

Along similar lines, some respondents who had worn T-shirts printed with explicit advocacy slogans and graphics discussed having motivations of peer education and persuasion simultaneously, often eliding any distinction between the two. This was particularly the case when wearing items geared toward raising awareness for issues that lie outside the mainstream, in typical reverse agenda-setting form. For instance, when Zachary was president of his high school's Darfur Awareness club, he arranged to have everyone in the group, including himself, wear "Stop Genocide in Sudan" shirts around the school. As he explained, "the whole point was definitely awareness . . . getting people at our school involved and understanding what the conflict was about." For Zachary, drawing attention to the facts of the Darfur conflict and promoting involvement in the political cause surrounding it were effectively one and the same. Such a dynamic was also noted by some who wore T-shirts promoting various aspects of environmentalism, another outside-the-mainstream cause in which the struggle to create an aware and informed public is seen as a necessary first step toward collective action. Brittney, for instance, described wearing a shirt that could be considered a kind of sartorial "teach-in": in addition to including the declarative advocacy slogan "Save Water," the design lists several alarming facts about water usage (e.g., "one ton of paper uses 7,000 gallons; one new car uses 39,000 gallons"). This combination of journalistic information and persuasive call to action is a typical example of advocacy journalism, only in concise T-shirt form. As Brittney wore it in public and around her peers, her roles as news spreader and cause promoter were vividly merged.

In addition to sharing explicit advocacy content for simultaneous education and persuasion, some respondents pointed to political satire media as providing them with opportunities to serve these dual functions for their peers in an entertaining and accessible way. For instance, many respondents who had shared a humorous web video on Twitter mocking the 2012 U.S. presidential candidate Mitt Romney emphasized the educational value of the clip, while at the same time embracing its clear anti-Republican bent. As noted previously, political satire media has increasingly become a source of political information for its audiences as well as a font of ideologically charged critique. The clip in question, which comically splices together a series of Romney's perceived gaffes from his speeches and television appearances (e.g., "I'm not concerned with the very poor") into a rhyming hip-hop song, typifies how contemporary political satire hybridizes news content and point-of-view commentary.[63] Lisa, who tweeted the video link to her followers, suggested that its comedic value made it useful for raising awareness about the election among her friends who do not tend to vote: "It was short enough and humorous enough that a lot of people could watch it . . . so I figured it would be a good way to get non-political people to learn political things." However, as she went on to explain, the specific lesson that she sought to impart with the video was decidedly partisan in nature: "A lot of my friends are from rural areas like Idaho where the news programs are definitely slanted towards the right, and they don't always get I think a full picture of what's going on in politics. So I try to show them some of the other side." In other words, Lisa saw the parody video's anti-Romney spin as balancing out the conservative bias of other partisan news content. Her comments thus underline how educating one's peers about politics in the contemporary partisan-heavy information environment may be less a matter of establishing agreed-upon truths than providing exposure to alternative ideological lenses on reality. In this context, political satire may be just as useful for accomplishing such a task as content formally designated as journalism.

As shown in these examples, negotiating the goals of spreading political knowledge and advancing specific political agendas presents more of a dilemma to some citizens than it does to others. Some who engage in news-spreading activities both online and offline make a point of distinguishing between these two projects. However, others who participate in the same kinds of practices freely move between educational and persuasive frameworks without reservations or caveats. Such a discrepancy brings to the surface the uneven tensions and anxieties that surround the incorporation of news-spreading practices into the arsenal of citizen marketing's "political ammunition."

At stake, so it seems, is the very identity of the journalistic enterprise. As the circulation of political information becomes increasingly driven by the selective interests of partisan-minded peers, does the notion of a disinterested knowledge base become obsolete? As the amount of political information explodes and attention becomes a scarce resource, will ideologically charged projects of raising awareness come to dominate how citizens learn about the world around them? Those who seek to uphold an idealized vision of journalism as a pursuit of common truths in the service of cooperative democracy would likely find these questions alarming. Indeed, the practice of sharing political information to advance specific causes, whether done with subtle care or more blatant fervor, can be recognized as contributing to the broader shift toward journalistic partisanship that threatens to upend the values of objectivity and professional news judgment often maintained by traditional gatekeepers. However, for those who cast a skeptical eye toward any elite claims of political objectivity, the growth of citizen-level participation in the circulation of news may be welcomed as long overdue.

In a sense, the participatory agenda-setting practices outlined in the above testimonies work from the assumption that *all* journalism is inherently colored by ideological bias, and that the citizen-level promotion of select issues is a direct response to the biases already bound up in the status quo. From Gabrielle's spot-lighting of stories that she finds are often ignored in the mainstream media to Lisa's attempts to provide a counterpoint to the dominant conservative voice of Fox News, acts of partisan-driven news spreading are often conceptualized as interventionist responses to a broader information environment that is seen as perennially falling short of lofty journalistic ideals of objectivity and impartiality. If mainstream news media ultimately function as the propaganda of elite pow-erbrokers, as Herman and Chomsky posit, then grassroots news spreading—at least that which diverges from elite agendas—serves as the agit-prop of myriad groups who seek to challenge these claims to truth.

Leaving aside the specific issue of sharing blatantly fake or false news, we can thus recognize how the citizen-level curation of political information broadly accelerates a postmodern vision of all truth as relative, partial, limited, and subjective. Depending on one's commitment to the upholding of a rationalist public sphere in the Enlightenment-bound tradition associated with figures like Habermas, such a view may be alternately feared or embraced. As Anderson argues, partisan-oriented information exchange is more aligned with Nancy Fraser's notion of subaltern counterpublics than with the traditional model of the public sphere.[64] In short, the counterpublics framework emphasizes the agency of marginalized and oppressed groups to counter dominant discourses through alternative networks of communication.[65] Under this perspective, the networked circulation of news content that draws attention to the political pri-orities of subaltern groups would be welcomed as an invaluable force of change that helps elevate the position of these groups within the broader society. There are no claims to political objectivity here, and this is precisely the point: rather than contribute to idealistic projects of journalism as rational consensus build-ing, counterpublics work to advance their own deeply invested agendas of social justice.

In the era of networked gatekeeping, these subaltern voices have newfound opportunities to bring their issues to the forefront and spread the news and infor-mation that serves their interests. This is especially the case for the host of pro-test movements that have largely mobilized online via platforms like Facebook and Twitter. From the Arab Spring in Tunisia and Egypt to the Indignados in Spain to Occupy and Black Lives Matter in the United States, groups represent-ing marginalized social interests have utilized peer-to-peer information circula-tion to make their grievances known to one another and to the outside world. As discussed earlier, these movements typically gain attention for their issues by making themselves the story, using networked digital media to self-document on-the-ground actions that become impossible for the world to ignore through the signal-boosting promotional labor of the critical periphery. Such hybridiza-tions of citizen journalism and citizen marketing thus open significant pathways for grassroots agenda setting on behalf of groups whose interests are poorly served by the journalistic status quo. Referring to these networks as "mass-self communication," Castells lauds their capacity to persuade broader publics in the service of contesting and decentering communication power.[66] With regard to Occupy, for instance, Castells posits that the most significant political success of the movement was its "impact on people's consciousness," particularly in terms of raising awareness of class struggles among the American polity.[67]

However, in the midst of this optimism over the digital grassroots, it is important to keep in mind that elites may also be empowered in an online information environment characterized by polarized one-sided information flows. As Gainous and Wagner remind us, one major risk associated with the shift from professionalized gatekeeping to social news filtering is the increased agenda-setting capacity of major political figures and organizations among their networks of supporters and followers.[68] Another issue that we have yet to explore in this chapter is the institutional power of for-profit news agencies, which capitalize on the promotional labor of citizen news spreaders to thrive in an information landscape dominated by social media distribution. In the closing section of this chapter, we will take a closer look at this emergent viral news ecosystem, which has become increasingly focused on the economic value of shaping content to maximize spreading and sharing behaviors, particularly among polarized networks of partisan audiences. As I argue, the resultant melding of commercially oriented soft news values with the agenda-setting aspirations of advocacy journalism poses significant challenges for citizens—as well as undoubtable opportunities—that closely mirror the dilemma of packaging in political marketing. As in the context of electoral promotion, citizen participation in the circulation of the emotion-laden and culturally resonant symbols that often dominate today's viral news cycles has the potential to both invigorate and enervate substantive engagement with structural political realities.

For-Profit Online News, Viral Cause Célèbres, and the Political Packaging Debate Redux

At the start of this chapter, we considered how the promotional labor of citizen news spreading is often focused on the aggregate production of publicity for strategically chosen individuals and groups in the broader information environment. Virtually overnight, a figure like Sandra Bland or Eric Garner can rise from obscurity to viral Internet fame as a result of networked information-sharing practices. As we have noted, however, the emergence of such cause célèbres is rarely, if ever, the result of purely grassroots amateur media participation. Rather, these processes are situated in the hybrid media system, which refers to the integration and interdependence of journalists, politicians, activist groups, and citizen networks in the production of political information cycles. Chadwick argues that in the hybrid media environment, power accrues to those who effectively navigate both old and new media logics: "actors create, tap, or steer information flows in ways that suit their goals and in ways that modify, enable, or disable others' agency, across a range of older and newer media settings."[69] In this context, the success of news professionals in the digital age hinges to no small degree on their mastery of the viral dynamics of peer-to-peer sharing, as well as the related ability to tap into existing networks of citizens whose media circulation activities align with organizational goals. Thus, as citizen news spreaders take to the Internet to raise awareness for select issues and create fame for cause célèbres, news agencies are presented with opportunities to take advantage of this activity to serve their own needs.

In the world of for-profit journalism, economic needs are perennially front and center, and this includes websites and blogs that are driven by partisan agendas. To be certain, such a partisan orientation is likely to reflect the ideological commitments of a news outlet's owners and operators (such as in the above-noted example of *Upworthy*, whose founder and chief executive has notable left-wing activism credentials). However, Tewskbury and Rittenberg point out that audience segmentation by political leaning is also good for the bottom line: "by tailoring their content to specific segments of the audiences, sites can package audiences for advertisers."[70] The authors further note that political divisions among audiences create some of the most valuable categories for advertiser targeting (alongside demographics such as age and gender) because they tend to be strongly correlated with patterns of lifestyle consumption.

However, catering news content to fit the political preferences of the readership is not the only way in which online journalism on the web is shaped by economic considerations. As Robert McChesney explains in his political economy of digital journalism, profit motives also lead online news outlets to prioritize the sheer audience appeal of content, which inevitably conflicts with traditional standards of professional news judgment. The result, he argues, is a pattern of soft news that has similarly characterized commercial journalism in older media formats, where "an increasing portion of news has gone over to inexpensive-to-cover entertainment, celebrity, gossip, crime, and lifestyle journalism."[71] As an example, McChesney cites an internal memo from the head of AOL, the owner of the *Huffington Post*, demanding that "all stories . . . are to be evaluated according to their 'profitability consideration,'" that is, their potential for web traffic and corresponding advertising revenue.[72]

In the era of Facebook, Twitter, and social media more generally, such traffic is increasingly driven by peer-to-peer sharing, underscoring how the profitability of digital news content has become greatly contingent on its potential for virality among citizen networks. According to 2015 Pew Research data, for instance, 63% of U.S. Facebook users and 63% of U.S. Twitter users report getting news from these platforms.[73] The question for commercial news agencies, then, is how to shape content for viral success. As the growing body of research on the psychology of viral sharing behavior suggests, the key is to elicit strong emotional reactions. For instance, a widely reported study by Jonah Berger and Katherine L. Milkman, which analyzed the social sharing of thousands of *New York Times* articles online, concludes that content that induces high levels of physiological arousal is most likely to be forwarded to peers; this includes both intense positive emotions such as awe and inspiration and intense negative emotions such as anger and anxiety.[74] In press coverage of the study, journalists were quick to link these results to the recent economic patterns of the viral news ecosystem, suggesting that the most lucrative sites have already figured out this formula and are using it to their advantage. A piece for the *New Yorker*, for instance, posits that Berger and Milkman's findings on the importance of arousal "go a long way toward explaining the success of websites like *Upworthy*, which . . . is known for using headlines designed to make you laugh, cry, or feel righteous anger."[75]

Such a focus on the emotional appeal of content would appear to push viral-minded news outlets toward the soft news values of entertainment and personality-driven drama over the traditional journalistic mission of "produc[ing] the necessary information for citizens to understand and participate effectively in their societies," as McChesney puts it.[76] However, for many of today's partisan news sites that take on an explicit issue advocacy role in much of their coverage to cater to their target audiences, the commercial motives of tapping into viral appeal go hand in hand with the more politically oriented concerns of agenda setting. In other words, these outlets package their ideologically inflected news for maximum consumption and peer-to-peer circulation, engaging in a process that resembles the emotion-grabbing and stylized tactics of political marketing. Indeed, the packaging dynamic is arguably a defining characteristic of the for-profit partisan news ecosystem, in which the quest for social media shares encourages the creation of content that both entertains groups of polarized audiences and conveys political viewpoints in ways that fuel their participatory agenda-setting. Here, the debate regarding political packaging that we have examined in previous chapters helps illuminate what is potentially at stake in such a formulation.

On the one hand, as sites like *Upworthy* seemingly uphold, wrapping political issues in emotionally arousing and popularly accessible forms and narratives enhances their capacity to impact the public agenda by eliding the boundaries between political and cultural spheres. On the other hand, as critical perspectives on trivialization suggest, the style-heavy packaging process risks emptying political content of its substance and thus weakening its potential for meaningful and consequential impact. Importantly, either scenario is wholly compatible with the economic incentives of for-profit journalism, since the shaping of news content for viral sharing promises to boost clicks and ad dollars. However, for partisan news outlets that aspire to affect political outcomes at the same time as they succeed financially, the packaging approach may not necessarily serve the former goal as well as it does the latter. Moreover, the citizen news spreaders who seek to meaningfully impact the public agenda appear to have more to lose from the proliferation of viral-ready content that risks trivializing the issues that they care about the most. Although their promotional labor in spreading this content to peers helps the commercial bottom lines of favored news outlets, whether or not this effectively functions to advance their favored causes on the broader political stage is altogether less certain.

The tensions between style and substance in the viral news ecosystem are particularly apparent in the phenomenon of Internet cause célèbres with which we began this chapter. As discussed previously, the act of producing publicity and fame for select individuals—achieved through a hybrid of professional and amateur information circulation practices—has the potential to create powerful symbols that personalize complex issues and resonate with publics on deep emotional levels. At the same time, however, when these symbolic figures become fodder for celebrity-like coverage that takes advantage of such resonances to maximize clicks, viral shares, and ad revenue, the result may be spectacle and sideshow that can distract from the substantive issues for which they are meant to raise awareness.

Both possibilities can be glimpsed in one of the most high-profile news stories of 2015, which involved the ascendance of the U.S. pharmaceutical executive Martin Shkreli to the position of "the most hated man on the Internet."[77] Shkreli became the subject of international attention, infamy, and loathing when his company raised the price of a drug used to treat HIV patients by 5,000%. Although the issue of unethical price gouging and lack of regulation in the pharmaceutical industry had been raised before in the press as well as in government, the dramatic nature of the Shkreli story promised to skyrocket it to the top of the U.S. public agenda and bring about significant policy change. For instance, shortly after the story first broke in September 2015, the presidential candidate Hillary Clinton announced via Twitter that "price gouging like this in the specialty drug market is outrageous. Tomorrow I'll lay out a plan to take it on;" this had the effect of causing a temporary crash in biotech stocks over industry fears of increased government regulation.[78]

However, Shkreli soon became far more than a mobilizing symbol for challenging the excesses of free-market capitalism in matters of public health and beyond. Rather, as a series of Shkreli's tweets showcasing his bombastic personality and lavish lifestyle created a viral sensation on the Internet[79] and earned him the derisive nickname "Pharma Bro," news reporting quickly reached celebrity-level proportions. Although Shkreli received substantial mainstream journalistic coverage, the intensive, paparazzi-like scrutiny was particularly prominent among left-oriented U.S. news sites and blogs,[80] where he was elevated to the level of popular culture villain through scores of castigatory articles documenting everything from his unsavory online dating activities[81] to his extravagant purchase of an expensive limited-edition hip-hop album.[82] Judging from the tone of Shkreli-related articles in left-wing online news outlets, strong emotional responses of anger and antipathy were major forces driving this coverage. *Raw Story*, for example, included him in posts like "Here Are 14 Douchebags We'd Love to See Disappear in 2016"[83] and "Here Are the 11 Most Punchable Faces of 2015."[84] This outrage turned into unbridled joy when Shkreli was subsequently arrested on securities fraud charges, which were unrelated to the drug-pricing controversy. As *Gawker* put it in a post entitled "Here's Your Martin Shkreli Perp Walk Schadenfreude Gallery," "let's all look at the bad man in chains . . . these pictures of the villain getting his comeuppance will soothe you."[85]

The Shkreli saga dramatically illustrates the tendency of online news outlets to magnify coverage of stories and personalities that have already achieved the status of viral fame among many social media users, which, as research suggests, is fueled by strong emotional responses. Indeed, many of these news articles focused on impassioned social media reactions to the story, such as *Salon's* "Twitter Nearly Drowns in Schadenfreude at News of 'Pharmo Bro' Martin Shkreli's Arrest Thursday," which compiled snarky user tweets like "Wait, Martin Shkreli is in jail? Are the criminals ok?"[86] In a for-profit digital news environment shaped by economic pressures to maximize clicks and shares, ramping up coverage of figures like Shkreli who have captured the attention and passion of key target audiences is clearly good for business. The barrage of content that both followed and lambasted Shkreli's every move once the initial drug-pricing story broke—regardless of its relevance for this specific policy issue—was undoubtedly connected to these financial considerations, as digital news outlets capitalized on their audiences' desire to broadcast their visceral condemnation of Shkreli to their peers.

Yet is it fair to label this commercially minded, sensationalistic, and celebrity-oriented coverage as soft news, the kind that scholars typically decry as distracting from the serious business of democracy-serving journalism? After all, it could be argued that every mention of Shkreli in the hybrid media environment, whether focused on his frivolous personal dramas or not, helped push the issue of pharmaceutical industry regulation further onto the public agenda. The act of making Shkreli famous as "the most hated man on the Internet," achieved through a combination of crowdsourced and professional gatekeeping practices, could very well have long-term "impact on people's consciousness," as Castells puts it.[87] This is because it creates a symbolic entry point to into a complicated policy matter at the accessible and affect-laden level of popular culture, which, as optimistic analysts of political marketing like Margaret Scammell argue, is a crucial dynamic for fostering democratic engagement more broadly.[88] However, the connection between the representative symbol and the broader structural issue may not necessarily be strengthened by coverage that follows a commercial soft news pattern of emphasizing sensational personal drama as a way of boosting audience interest and peer-to-peer sharing. Perhaps the biggest potential pitfall of the Shkreli viral news cycle would be to frame the solution to the political problem he has presented in wholly individualized terms—for instance, by reveling in the "comeuppance" of his arrest for an unrelated offense. Although this may offer a satisfying sense of closure for those who contributed promotional labor to making Shkreli an infamous cause célèbre, it leaves the larger policy issues that surround his story virtually untouched.

The journalists and citizen news spreaders who together brought Shkreli to the height of Internet notoriety fostered an impassioned enthusiasm for the story, but was this a route to material change or merely a simulation thereof? In her discussion of affective publics in digital milieus, Papacharissi highlights this dilemma when she addresses how emotionally fueled responses to news via social media can draw citizens into political affairs in powerful ways, yet cannot guarantee movement toward substantive political outcomes: "affect is intensity, and intensity can both reflexively drive forward and entrap in constantly regenerated feedback loops."[89] Here, Papacharissi points to a forking path of mobilization and trivialization that would appear to characterize the viral news ecosystem as a whole. In the case of the Shkreli saga, for instance, the intensive emotional responses that drove its virality ran the risk of slipping into an endless loop of public shaming and ridicule aimed at a single individual, even while it held the promise of activating citizens around a policy issue that may have previously been viewed as overly abstract or obtuse. Furthermore, Papacharissi warns that "affective mechanisms increase awareness of an issue and in so doing amplify the intensity of that awareness. They do not inherently enhance understanding of a problem, deepen one's level of knowledge on a particular issue, or lead to thick forms of engagement with public affairs."[90] As the professional journalistic enterprise becomes increasingly geared toward exploiting these affective mechanisms for financial gain in the era of social media news spreading, this point is important to bear in mind. The chase to maximize emotionally driven viral shares does not necessarily preclude substantive knowledge exchange, yet it may create more traps into which citizens can fall as the lines between hard and soft news are increasingly blurred in a for-profit journalism landscape.

What, then, is the way forward for citizens who wish to make meaning-ful contributions to participatory agenda setting through their promotion of select news stories? As I have suggested, commercial news outlets cannot be relied on in and of themselves to fasten connections between attention-grab-bing political symbols like viral cause célèbres and broader structural issues, since they have a vested financial interest in attracting the momentary engage-ment of their target audiences, regardless of the long-term political conse-quences. Rather, the burden appears to fall largely on citizen news spreaders themselves, who must consider not only what stories to help publicize as a means of raising awareness for favored issues, but also which framings, or ver-sions, of these broader narratives are ultimately best positioned to advance their agenda-setting goals. As I discuss in detail in the following chapter, this requires developing a critical literacy of citizen marketing practice that, among other things, pauses to reflect on the hermeneutic processes of meaning mak-ing that link symbolic political content to larger structural realities. In other words, although many citizens may be eager to boost the signal of events and narratives that align with their political priorities, they stand to benefit from sorting out which frequencies of the signal hold the greatest potential for car-rying the deeper political meanings they wish to get across, and which fre-quencies have the greatest risk of becoming stuck in loops of distractive noise. No single formula can be universally applied to make such determinations, and it would be foolish to suggest otherwise. What matters, ultimately, is the meaning making that emerges from the circulation of political information, which is always by definition contextual and contingent.

Conclusion: In the Trenches of Information Warfare?

This chapter has argued that the act of selectively forwarding political news and information to boost its public profile has an uneasy, yet inextricable, relation-ship to the instrumental persuasion dynamic of the citizen marketer approach. As vividly illustrated in crowdsourced actions to produce fame and infamy for certain individuals, groups, and narratives in the broader information environ-ment, citizen news spreaders take on a publicist-like role that mirrors the stra-tegic dissemination of "interested fact and opinion" (in the words of Harold Lasswell)[91] bound to traditions of propaganda and its latter-day offspring, public relations. However, considering that professional journalism as a whole has been accused of manufacturing consent for elite agendas, citizen-level intervention in political information flows holds the promise of disrupting this hegemony and advancing the agendas and priorities of a far broader range of publics and counterpublics. The result, it would appear, is a strongly agonistic scene of information warfare, far removed from the deliberative ideals of disinterested knowledge exchange that have long characterized the high-minded aspirations of the journalistic enterprise.

As the above testimonies of citizen news spreaders suggest, the conflict between the goals of civic education and public persuasion in the exchange of political knowledge is not merely a matter of academic debate. Rather, it is a tension that is continually negotiated at the granular level of everyday cultural practice. At a time when political information itself has grown more and more ideologically charged across a variety of formats and genres, from partisan news channels, websites, and satire programs to activist citizen journalism and institutional one-sided information flows on social media, this tension appears to be increasingly unavoidable. Depending on one's comfort level with the notion of journalistic information as a tool of agit-prop, the hybridization of knowledge exchange and peer persuasion may be viewed as either unsettling or empowering.

In addition to grappling with the potential exacerbation of journalistic partisanship through the agenda-driven—and agenda-driving—practice of networked gatekeeping, citizens who participate in the selective promotion of news flows must further contend with the economic dynamics of a largely for-profit digital news ecosystem that is incentivized to maximize viral audience appeal above and beyond any other concerns. Although this poses notable risks of trivialization that should not be ignored, it also creates opportunities to take strategic advantage of the affective intensities of political symbols, which, as Papacharissi argues, has the potential power to disrupt broader social narratives and present "semantic challenges" to established ways of thinking.[92] This is indeed a central dilemma for all citizen marketing practices that center on symbolic action, including those that focus on promoting informational awareness as a first step toward long-term change. As noted earlier, the citizen marketer approach tends to operate under what Zuckerman calls the "cultural theory of change"—that is, one that envisions efficacious political action as a process of shifting cultural mindsets in a diffuse and often highly subtle fashion.[93] Although some tangible markers may exist, cultural change in the broadest sense may elude many forms of concrete measurement.

It is this fundamental indeterminacy of cultural processes, perhaps more than any other factor, which helps account for why the increasing emphasis on media-based symbolic action within the repertoire of political participation is so thoroughly controversial. In the next chapter, we will explore this controversy in detail, reexamining the charges of slacktivism that are often targeted at participatory media efforts organized around the peer-to-peer production of symbolic visibility, momentum, and awareness. Building on the questions of broader political consequences that surround the slacktivism debate, we will revisit the multiple ramifications of the citizen marketer approach that hold both promises and challenges for the future of democratic life. Then, we will sketch a model for developing a critical literacy of the citizen marketer that can potentially help both its practitioners and its critics to navigate its complex topographies.

6

Toward a Critical Literacy of
the Citizen Marketer Approach

Up the Ladder, Down the Ladder,
or No Ladder at All?

In the preceding chapters, we saw how the peer persuasion logics of the citizen marketer approach operate within a wide array of contemporary political phenomena, from the advocacy efforts of identity-centered social movements to the promotion of election campaigns to agenda-setting flows of news and information. Now that we have a firmer sense of what citizen marketer activity hopes to accomplish, how it hopes to accomplish it, and what key issues and tensions it raises for civic and social life, we are in a better position to address its broader consequences. To do so, this concluding chapter focuses on how we might develop a critical literacy of the citizen marketer approach that can evaluate the nuances of both its opportunities and its risks and situate it more firmly within its surrounding political, social, and cultural contexts. Ultimately, my goal is not to argue whether the citizen marketer approach is a priori good or bad for democracy in a normative sense, or whether it should be embraced or rejected by citizens who seek to make a difference in a media-saturated political world. Rather, the purpose of developing a critical literacy is to obtain the necessary analytical tools to arrive at one's own answers to this difficult set of questions. In addition, for those who are drawn to the promise of citizen marketing's persuasive power, a critical literacy can potentially provide direction for harnessing this power more effectively as well as more responsibly, offering a compass of sorts for navigating its numerous challenges.

As a first step toward developing this critical literacy, it is necessary to consider the relative role of citizen marketing practice in the broader field of citizen-level political action, leading us back to the slacktivism controversy with which we began this book. In an era of increasing media participation, buoyed by the explosion of peer-to-peer social sharing on digital networks, citizen marketing practices of all kinds appear to be growing and intensifying across the political landscape. Considering that they have become so popular and so commonplace, how might they be changing the nature of political participation more broadly, and what are the consequences for the health of democracy? These are the questions that inevitably seem to dominate any conversation of the subject. Those who pose it often come from a place of deep skepticism toward symbolic, media-based forms of political action, particularly those that exist outside the boundaries of the "classic" collective action repertoire established in the 19th century—participation in organized rallies, marches, demonstrations, and the like.[1]

By contrast, citizen marketing tends to operate under what Bennett and Segerberg refer to as the newer logic of connective action.[2] Although symbolic media-spreading practices may be highly coordinated—from unified profile picture–changing and T-shirt-wearing campaigns to the en masse selective forwarding of advocacy and news content—they are also highly individualistic in nature. In their discussion of the changing dynamics of organized activism, Bruce Bimber, Andrew Flanagin, and Cynthia Stohl emphasize how digital technologies expand the agency of citizens to adopt varying styles of participation in political organizations, and that an "individualist" style, which takes an entrepreneurial stance to achieving collective goals, may be particularly well supported in the digital communication environment.[3] As the authors note, this shift to newer and less traditionally defined relationships between citizens and formal political organizations may be "disorienting . . . as new technologies and new ways of belonging continue to multiply in advance of the development of new norms and customs about how the practices of collective action should work."[4] This lack of established norms may help account for why the citizen marketer approach—which closely aligns with the individualist participatory style in the sense that it focuses on self-expressive acts aimed at one's immediate peer networks—has become an object of panic and uncertainty in both activism circles and scholarly debates. Dahlgren, for example, worries whether the "solo sphere" of individualized online political expression will foster an altogether weaker, "lukewarm" form of activism because it has a lower affective intensity in comparison to the high intensity of traditional collective actions such as on-the-ground demonstrations.[5]

Indeed, one of the core concerns regarding the growth of individually focused and media-centered forms of political action is that they seemingly violate the well-established norm that affecting real political change is supposed to be *hard*. By contrast, symbolic mediated action is remarkably low cost and easy to perform. Within digital contexts, it often takes little more than a few clicks of a button. Malcolm Galdwell, one of the fiercest and most high-profile critics of slacktivism, argues that political activism that can truly change the world demands a level of collective sacrifice that is only possible through traditional on-the-ground organizing. In a widely circulated *New Yorker* piece decrying social media–based political movements, Gladwell states that "Facebook activism succeeds not by motivating people to make a real sacrifice but by motivating them to do the things that people do when they are not motivated enough to make a real sacrifice."[6] This distinction that is frequently drawn between the ostensibly difficult work of organizational-based participation and the relatively facile and low-cost work of individualistic, media-based participation forms the foundation of what Henrik Christensen calls the substitution thesis: presented with a choice of these two forms of activity as equally valid options, citizens would naturally gravitate to the latter in lieu of the former.[7] After all, why do something challenging when there exists a far easier alternative that promises the same results of advancing one's interests in the broader political world?

As Christensen explains, the critique of slacktivism is "aimed at low effort activities that are considered incapable of furthering political goals as effectively as traditional forms of participation. Wearing badges is not enough, and neither is changing your profile picture on your Facebook account. . . . The slacktivists are seen as unwilling to get their hands dirty."[8] Morozov, who has popularized this line of thought, argues that "the real issue here is whether the mere availability of the 'slacktivist' option is likely to push those who in the past might have confronted the regime in person with demonstrations, leaflets, and labor organizing to embrace the Facebook option and join a gazillion online issue groups instead."[9] Such anxieties about the use of symbolic action as a substitute for more ostensibly effective forms of organized collective action are indeed widespread. When conducting interviews for this project, I heard several invocations of the substitution thesis from respondents, particularly those who are also engaged in organized activism and fear that its political force will be undercut by the growth of the citizen marketer approach. Regarding the red equal sign Facebook profile picture campaign for marriage equality discussed in Chapter 3, Riley offered a typical slacktivism critique: "I was really kind of peeved that there wasn't actually any action being taken. These are folks who are activists in a sense that, you know, they'll post about things on the Internet, but they won't be doing anything else . . . it's a thing that people think they can do and then wash their hands of the issue."

As Riley's remarks underscore, the biggest concern of slacktivism critics is not just that symbolic action may be ineffective in producing desired political change, but also that it may be counterproductive by providing a self-satisfying—yet impotent—alternative to traditional collective action. In their study of British trade union activists' use of social media, Fenton and Barassi note that many leaders of these organizations concurred with this line of thought, "believ[ing] that political participation on social networking sites distorted people's understanding of collective action, by reinforcing the idea that simply joining a Facebook Group was enough."[10] From this perspective, citizen marketing amounts to a form of disempowerment, leading citizens astray from the tried-and-true methods of organized activism and thus diminishing their capacity to meaningfully advance their political interests. Furthermore, as we saw in Chapter 1, the fear of individual self-fulfillment being prioritized above all else in one's political activities has been tied to the critique of the consumerist model of citizenship more broadly, which is seen as a result of pervasive neoliberal market logics colonizing the space of the political.[11] Such a perspective is seemingly buoyed by the fact that today's symbolic political actions are so often performed via commercial and corporate digital platforms like Facebook and Twitter that are wholly situated within these capitalist logics, and that foster the personalized consumption and sharing of political media to serve their own economic needs.

However, the slacktivism view of disempowerment through self-satisfaction contrasts sharply with the culturalist agency-building framework outlined in Chapter 1, which tends to position mediated political interaction among citizens as a foundational step toward more traditional and organizationally based forms of political participation in the future. For instance, Giovanna Mascheroni describes how informal engagement with a Facebook group created by the Italian anti-Berlusconi organization Popolo Viola served as a key mobilizing force for participation in the group's later rallies and demonstrations. Mascheroni argues that this path to mobilization was powered in part by the "perceived common identity symbolized by the purple color" that was fostered by the Facebook group.[12] Here, symbolic media-based practices are conversely framed as a form of citizen empowerment, in the sense that they motivate people to do more at the organized level by fostering identification with a group, movement, or cause.

This notion of lower-cost symbolic media participation pushing people toward—rather than away from—higher-cost forms of organizational participation is indeed a key logic behind large-scale advocacy groups' deployment of citizen marketing tactics. For instance, Anastasia Khoo, the marketing director for the HRC organization and creator of the red equal sign campaign, explains that for the many heterosexuals who changed their Facebook profile pictures to the marriage equality advocacy image, "this was the first step to becoming a straight ally, and now it's our job as advocates to help bring these individuals *up the ladder of engagement* so that the next time they receive an email from us, asking them to contact their legislator, they might do so" [italics mine].[13] In other words, the HRC deliberately encouraged this symbolic action on Facebook for the purpose of organizational outreach, hoping that the mass adoption of the image as a badge of identity would foster new ties to the organized LGBT movement and inspire higher-cost participation later. The metaphor of moving people "up the ladder of engagement" is essentially the reverse of the substitution thesis, which could be thought of in parallel terms as a model of moving people *down* this same ladder.[14]

Thus, we have at our disposal two opposing and conflicting theories regarding the relationship between citizen marketing practices and traditional organized political participation: one imagines a misguided self-satisfaction leading to disengagement with more effective forms of action, whereas the other imagines a strengthened personal identification as inspiring this very sort of activity. So, which is it? Regarding the former, Christensen notes that existing empirical research on civic engagement and Internet use shows "no evidence that Internet activities are damaging civic engagement by replacing more effective forms of participation."[15] Likewise, not a single respondent for this project indicated that they would avoid organized political participation in the future because they felt that they had already done enough for their favored cause by circulating media content. Of course, this finding alone does not rule out the substitution thesis entirely, and it remains a distinct possibility for at least some portion of the citizenry. Yet at this stage, it appears to be less a theory supported by concrete evidence than an amorphous fear of how unknown others will behave.

Regarding the converse up-the-ladder-of-engagement model, we have seen in previous chapters how the fostering of political identity championed by the culturalist framework of identity and agency building is bound up in many of the logics that fuel citizen marketing practice. For instance, in Chapter 3, we noted how symbolic visibility campaigns among gay and lesbian citizens and allies are often seen as serving a key intracommunity role (in addition to an external public advocacy role), shoring up a sense of collective identity and common purpose in ways that can potentially build the ranks of a political movement. Although this model may be particularly pertinent for issues of identity politics, the idea of stepping out from the shadows with symbolic markers to strengthen communal identification and foster deeper engagement may be relevant for a wide range of constituencies. As discussed in Chapter 4, this includes the field of electoral politics, where a focus on building energy for campaigns by encouraging displays of party identification and candidate veneration among the like-minded may be linked to attempts to boost organizational engagement (such as donations or volunteer work). However, it remains unclear at this stage whether participation in various sorts of citizen marketing efforts actually pushes people up the ladder of engagement in the ways that they might be intended by organizations such as the HRC. On the one hand, we can assume that organizational participation requires a strong level of identification with the group in question (as well as the broader cause that it represents) and that this identification may potentially be cultivated by symbolic acts such as public self-labeling. On the other hand, it is also obvious that wearing a political organization's T-shirt or posting its logo or hashtag by no means guarantees any further organizational involvement. This is precisely the concern of the slacktivism critics—*feeling* like a member of a political group or movement is not the same as *working* on its behalf.

However, the fundamental problem with the ongoing debate between up-the-ladder optimists and down-the-ladder pessimists is that they both evaluate the efficacy of symbolic political action quite narrowly in terms of how it impacts other forms of political participation. In both cases, the persuasive dimension of these practices—which, as illustrated in the accounts presented in this book, can be a primary motivating force for many of its participants—is simply ignored or dismissed. Furthermore, both theories assume that there *is* in fact a causal relationship between symbolic media participation and more traditional forms of organizational political participation, with the only difference being whether the former is seen as either encouraging or discouraging the latter. These theories thus neglect to consider how citizens understand the potentially divergent roles of these categories of politically oriented activity, as well as how they may not necessarily be mutually exclusive or in competition with one another. Indeed, a third possibility remains: namely, that there may be no causal relationship between these kinds of activities at all, at least for a certain portion of the citizenry. Such a theory would therefore require us to reject the ladder metaphor altogether.

What would this "no-ladder" theory look like in practical terms? First, it would suggest that citizens who are already organizationally involved would likely remain so, despite adding new forms of symbolic labor to their repertoires of political action. This is, of course, in direct contrast to the substitution thesis, which suggests that citizens will lose interest in more traditional forms of organized political participation after being offered easier, low-cost, media-based alternatives. In interviews for this project, respondents who described themselves as committed activists often made a point of emphasizing that their symbolic actions online and elsewhere must be complemented by more traditional organizational work. For instance, Peter, who noted engaging in many organized actions in addition to changing his profile picture to the red equal sign on Facebook, remarked that this activity "doesn't take away from other things. To me it's not an either or thing." Similarly, David commented that for him and his husband, participating in this public advocacy campaign for marriage equality is "one of many things we should be doing. Like in my personal case, you know, it's not the only thing that we were doing as a couple. We also were donating." As these anecdotal examples suggest, a commitment-enervating sense of self-satisfaction is far from a universal experience for citizens who participate in symbolic practices that align with the citizen marketer approach.

Although this theory positing no causal relationship between media-based action and organizational participation suggests that those who are already committed to the latter remain so despite their engagement in the former, what does it have to say about those who take up symbolic action as their only contribution to a cause? In contrast to both the substitution thesis and the up-the-ladder-of-engagement model, some respondents suggested that citizen marketing practices expand the circle of political participation to include those who are unlikely to ever become active at an organized level. As Andrew put it regarding the slacktivism critiques of the red equal sign campaign, "I think that a lot of people who maybe engaged in it wouldn't have done anything at all otherwise. And so to say 'oh, you know, it's not enough, it's not good enough' . . . they're not under an obligation to do anything at all." Kris, another participant in the campaign, similarly argued that "even if the vast majority of people just change their picture, that's still better than nothing. . . . That's still spreading awareness, and that's still something."

This notion of expanding participation to a new swath of public was also suggested by some respondents who noted that they personally would not consider taking action beyond the symbolic level. For instance, Evelyn remarked with regard to the issue of marriage equality, "I would never sign a petition, I would never march in the streets, but I could put up this equal sign and show everybody that I agree." Evelyn's symbolic action on her Facebook page thus marked the first time that she was ever involved in the issue, and although this may have been the furthest she was willing to go, her comments indicate that she would have remained wholly inactive otherwise. For those like Evelyn who had no previous involvement in the marriage equality movement and had no plans to engage at an organized level in the future, the symbolic connective action on Facebook thus provided an opportunity to at least make a microlevel contribution to an aggregate project of media visibility and public advocacy. Such an outcome would appear to fall short of the hopes of the HRC's Khoo and other movement organizers to push these new entrants up the ladder of engagement in the future. However, if there is indeed any political value to be found in persuasion-focused citizen marketing efforts such as self-labeling visibility campaigns, then this would suggest that the movement would potentially benefit by broadening its scope to include purely symbolic participants as well as those who are also organizationally engaged.

Thus, this third theory sets aside the ladder of engagement as the prevailing framework for conceptualizing the political efficacy of citizen marketing and shifts the focus back to its persuasive dimension, which is seen as having potential political value in and of itself. Following this reasoning, those who participate in a political movement or campaign purely at the symbolic level would not be slacktivists who are derelict in their responsibilities to become more involved. Rather, they would constitute an additional subgroup or layer, whose participation in projects of peer persuasion may potentially supplement—and, crucially, not detract from—the work of formally organized activists who also pursue the

same political goals with more traditional tactics. Such a position recalls the concept of the critical periphery of protest movements discussed in Chapter 5, who
serve a key role in amplifying the reach of protest messages by circulating them
on platforms like Twitter.[16] In fact, Sandra Gonzelez-Bailon, a member of the
research team that produced the critical periphery study, draws a crucial connection to the peer persuasion framework of citizen marketing in explaining the
group's findings: "If you want a product to go viral or you want a protest to grow,
you need that influential core, but you also need the periphery echoing them.
Peripheral users are not 'slacktivists.' They are quintessential to understand why
products go viral or protests go big."[17] In other words, this peripheral layer of
non-core, symbolic-only participants serves a message promotion function that
complements the high-cost, high-intensity actions of activists who share common political goals. Each layer is understood as having a valued role to play,
making concerns over "moving up" or "moving down" between them beside
the point.

This theory positing no ladder can be summarized thus: *citizens who are
organizationally active, as well as organizationally inactive, will tend to stay in place
in terms of their level of organizational commitment when engaging in symbolic
political actions via participatory media platforms.* Importantly, this accounts for
the likely possibility that organizationally engaged activists constitute a significant portion of the overall ranks of the citizen marketer, as new forms of
media participation are embraced as additional tools of activism rather than
as substitutes. This is in keeping with the testimony of the many organizationally involved citizens featured in this book who strongly emphasized their commitment to both media-based and organizationally based forms of political
participation. Furthermore, it is also in keeping with the long history of symbolic political action within both electoral and social movement contexts. As
discussed in Chapter 2, members of political organizations have for centuries
incorporated the display of symbolic artifacts (banners, ribbons, sashes, buttons, placards, etc.) into the standard repertoire of organized collective action
such as rallies and demonstrations. However, although organizationally active
citizens may contribute a considerable amount of symbolic promotional labor
as part of their overall repertoire of political participation, the popular uptake
of low-cost expressive formats enabled by digital technologies would appear to
expand this labor to a much larger group of organizationally inactive citizens,
thus magnifying its overall impact.

To conclude our discussion of citizen marketing's relationship to traditional
forms of organizational political participation, it is important to emphasize
that there is still little available evidence to either support or challenge the no-
ladder hypothesis, and future research is needed to explore it more thoroughly.
However, even simply recognizing it is a possibility helps to move the discussion

about citizen marketing beyond the strictures of the slacktivism debate, which tends to set up a simplistic binary of media-based symbolic action as being either good or bad for traditional activism. Indeed, focusing only on citizen marketing's possible ramifications for other forms of political participation conceals many of the more complex and pressing questions that surround this distinct phenomenon in its various permutations.

These questions, as signaled above, begin with the potential efficacy of citizen marketing as a distinct type of political action: in what ways might this media-oriented form of persuasive activity actually make a substantive impact in what Castells calls the battle over the minds?[18] In what ways might it shift public opinion in favor of particular positions, or strengthen existing positions, in ways that are desired by its practitioners? In the previous three chapters, we have seen how those who participate in citizen marketing practices tend to organize their efforts around three broadly defined goals: boosting the visibility of certain identity groups as a means of shifting perceptions of political reality and norms of social acceptability, modeling public enthusiasm and affect as a means of fostering momentum in partisan conflicts such as democratic elections, and increasing informational awareness about favored issues as a means of strategically reshaping the political priorities of the public. However, as we have also seen, the potential repercussions and consequences of these approaches span far beyond the issue of whether they are effective as tactics of political persuasion. Indeed, for each of these tactics, we have made note of their potential benefits in terms of expanding democratic participation, as well as their potential risks in terms of posing threats to deeply ingrained democratic ideals. At this point, it is productive to revisit and reassess these various critiques to gain a more robust perspective on the trade-offs of the citizen marketing approach. In the process, I will detail the concept of light politics, first introduced in Chapter 4, as a means of capturing these paradoxical dynamics in succinct form.

Before we begin, however, it is important to emphasize the following point: like all marketing, grassroots and otherwise, citizen marketing can—but by no means always necessarily will—have certain desired effects on the attitudes and behaviors of those who are exposed to it. As Michael Schudson argues in *Advertising, the Uneasy Persuasion*, we are right to be skeptical of the promotional industries' ability to manipulate the public mind in a mechanical, all-powerful fashion, yet this does not mean that their efforts to persuade are always by definition impotent or without purchase.[19] Likewise, depending on particular sets of conditions—including, but not limited to, the specific goals of any one promotional campaign or effort—participatory political marketing does have the potential to influence its audiences (as well as contribute to a range of other social and cultural consequences). Denying this would be akin to denying the capacity of political advertising to have any effect on the fortunes of electoral candidates, or of media messages and images more generally to have a broader ideological impact on its audiences. Taking a more nuanced approach to the tricky question of media effects, therefore, allows us to think more deeply and carefully about the range of possible ramifications, both intentional and unintentional, of citizen marketing as a growing phenomenon on the political scene.

Light Politics, or the Trade-offs of Citizen Marketing

As explored in previous chapters, the dynamic of sacrificing complexity for simplicity, or depth for shiny surface, appears to be endemic to the citizen marketing approach as a whole. To connect citizens together in symbolic political displays via accessible peer-to-peer platforms, some of the more nuanced and complicated aspects of the issues involved must inevitably be diminished or downplayed. This point resembles a familiar critique of political marketing more broadly—that is, the threat of trivialization, or the "dumbing down" of political issues in the process of packaging them as appealing products for consumption. However, within the context of the citizen marketer approach, it is important to recognize how the same dynamics that seemingly trivialize the political can also expand their potential for widespread participation and circulation.

Specifically, we have seen how the intermixing of political expression and popular forms of cultural expression—a phenomenon that, as discussed in Chapter 2, has been steadily building steam in advanced Western democracies since the countercultural upheaval of the 1960s—elides the boundaries between the formal sphere of politics and informal spaces of everyday life. From printed slogan T-shirts and bumper stickers that casually interject campaign messages into quotidian visual landscapes to politically charged posts and links that circulate within the leisure spaces of digital social media, citizen marketing practices infuse the political into more immediately accessible venues of popular culture. This process can be understood as an outgrowth or extension of the culturalization of political communication more generally, a concept that numerous scholars have observed through the analysis of everything from political satire television[20] to the role of MTV in youth-oriented get-out-the-vote campaigns.[21] As a multiplying array of participatory media platforms allow more citizens to enter the orbit of a deeply culturalized political communication sphere, the inherent popular appeal of this style of discourse can be recognized as a driver of mass participation. In other words, a more fundamentally entertaining and accessible version of political communication holds the promise of expanding involvement on the part of everyday citizens.

But at what price? Surely, something of great importance must be lost as citizen-level political discourse becomes increasingly inscribed within the confines of snappy memes and viral videos, of pithy slogans, headlines, and hashtags that stream with nearly effortless ease across both online and offline spaces of popular culture. At least that is what the multitude of critics who warn of political trivialization would urge us to consider. However, rather than take sides in an intractable debate over the normative value of the ever-growing politics–culture nexus, it is more productive to step back and trace the contours of the citizen marketing approach in a way that brings to the surface both its potentially liberatory and its disconcerting elements.

To this end, I offer the term light politics, which is intended to have a double meaning. On the one hand, the political content forwarded via citizen marketing practices can be described as light, as in reductive, simplified, or lacking gravitas. For instance, it is difficult to fit much of a sustained political argument into the space of a Facebook profile picture, graphic T-shirt, or 140-character tweet, or to advance an in-depth and well-rounded knowledge of the world with a few selectively chosen bursts of digital data. On the other hand, this content can be thought of as light in the sense that it can easily and casually travel across a culture, circulating political ideas and information (however briefly sketched) into the spaces and places of everyday life. These two qualities are indeed interrelated, since the lightness of content allows for lightness of movement. The light politics concept thus helps us to envision the complementary give and take of citizen marketing practices—the trade-offs incurred when sacrificing some of the substantive depth of political discourse to heighten its ability to circulate among a wider swath of the public and make an impact in a highly competitive attention economy.

For instance, in Chapter 3, we examined how citizens use the self-labeling capacities of participatory media platforms to announce their political identities on the public stage, particularly within contexts in which the presence of these identities had been previously invisible, unknown, or obscured. The case of gay and lesbian visibility campaigns, involving practices such as the symbolic display of self-disclosing T-shirts and social media profile pictures, alerts us to the capacity of peer-to-peer grassroots media to expand the visibility of marginalized groups in ways that top-down forms of mass media representation perhaps never could. At the same time, however, these coordinated displays of identity have a tendency to embrace easily legible symbolic labels that, although making for catchy and compelling marketing, also risk promulgating a simplified and flattened version of identity that can be reductive or exclusionary. The red equal sign profile picture campaign, for example, appeared to trade off diversity for group cohesion: in the process of presenting a forceful, united image of LGBT and ally identity to the broader public as an instance of viral mimesis, the campaign also left some members of the community who did not identify with the HRC organization and its equality rhetoric feeling hidden and left out—an ironic reversal of visibility. At the same time, the reflexive critiques of this campaign, which too circulated on platforms like Facebook via grassroots peer-to-peer efforts, provided at least some limited opportunities for LGBT citizens to address these absences and further expand the mediated representation of the community. Thus, the flattening of identity into the symbolic packets of light politics may be an unavoidable consequence of media visibility campaigns organized around the logic of meme replication, yet the widespread cultural circulation of these symbols may serve as the beginning of a conversation rather than an endpoint.

In Chapter 4, we saw how the fundamental tension of light politics also figures prominently in participatory electioneering efforts within the institutional political sphere. In modern Western democratic systems like that of the United States, the threat of diminishing citizen engagement has been met with efforts to transform partisan supporters into impassioned and evangelistic fans, appropriating culturally situated models of celebrity worship and brand identification to encourage peer-to-peer candidate promotion and "buzz." Through a variety of top-down and bottom-up efforts, members of partisan political factions work to perform grassroots enthusiasm and energy for their side in a way that they hope will galvanize the like-minded, seduce the undecided, and perhaps even win a few converts along the way. As participation in electoral campaigning increasingly becomes a site of popular culture engagement—akin to spreading the word to one's peers about a beloved object of fandom or gleefully jeering a hated sports team at the stadium in the hopes of tripping it up—we can presume that significant new pathways will open for groups of citizens, particularly young people, to enter the domain of formal institutional politics for the first time. We have already glimpsed what this process can look like in the campaigns of figures such as Barack Obama and Bernie Sanders in the United States and Jeremy Corbyn in the United Kingdom, where the widespread circulation of cultural artifacts such as humorous Internet memes, viral videos, and trending hashtags helped create a peer-to-peer promotional apparatus that significantly expanded these campaigns' voter outreach efforts.

Although the conflation of partisan foot-soldier and pop culture fan may hold the promise of democratizing mediated discourse surrounding elections and institutional politics more broadly, it also presents clear challenges to long held ideals of what "good" democracy should look like. In addition to potentially heightening affective polarization and political tribalism by reinforcing cultural divides, it also risks exacerbating the trend of elections as popularity contests between image and personality-oriented celebrity candidates—a process that has been in motion for many years in Western mass democracies and is currently expanding around the globe. Indeed, this trend appears to be gaining a newfound intensity at a time when peer recommendations on social media platforms have become a core logic of political marketing strategy. The emergent model of candidate promotion powered by lay supporters acting as grassroots intermediaries for favored political brands seems to almost guarantee an increasing emphasis on brand logics in the political realm more broadly—the emphasis on emotional forms of identification, the predominance of images as markers of cultural lifestyles, etc.

For those who fear the triumph of style over substance in a media-centered political world—following in the tradition of Walter Benjamin's warning of aestheticized politics as underwriting new forms of hegemonic control[22]—these developments are likely taken as quite worrisome. More than simply trivializing the political as a kind of *politics lite*, they seemingly threaten to deepen the power of elites to manipulate and hypnotize the citizenry with "all consuming images" (to borrow a phrase from the leading trivialization critic Stuart Ewen[23]), while distracting from the actual substance of issues. Furthermore, such concerns over elite image making and the strategic cultivation of candidate fandom dovetail with Stromer-Galley's critique of controlled interactivity in digital campaigning, where supporters are positioned as conduits for spreading campaign messages through the design of online tools and apps that systematize processes of political brand evangelism. However, it is also important to recognize how citizen marketing practices abound with a plethora of "unofficial" grassroots expressions that often overshadow official institutional attempts to shape peer-to-peer promotion, and how this complicates critiques of elite control. Moreover, we have noted how the co-creation of campaigns and candidate brands through participatory processes of symbolic circulation holds the promise of not only helping citizens attain a modicum of power and influence over a field of political marketing that largely pours down from above, but also enabling them to reshape this field in ways that more meaningfully resonate on affective and cultural levels.

In Chapter 5, we saw how citizens may also be empowered in the broader media landscape as they take on the role of networked gatekeepers that elevate the public profile of select issues that represent their invested political priorities.[24] By promoting certain pieces of information that comport with one's political interests while avoiding others that do not, citizen news spreaders act in a sense as publicity agents for their favored partisan figures, groups, causes, and movements. In the same way that public relations has been seen as a less serious and comprehensive (and perhaps even dubious) source of information because of its calculated slant on behalf of certain vested interests, the selective forwarding of political information at the citizen level can similarly be criticized as light—that is, lacking the full substance and gravitas of professionalized journalism. Although the forwarded bits of news content themselves may be taken from relatively thorough and professional sources, their recontextualization as political ammunition (to borrow Anderson's term[25]) by those who have an interest in promoting certain political priorities threatens to advance only limited and partial knowledge of a complex political world.

However, as the light politics concept suggests, this dynamic of reduction and simplification also creates opportunities for expanded popular participation in the agenda-setting process. It would be unrealistic to expect amateur news spreaders to serve as comprehensive intermediaries of political information in the vein of so-called newspapers of record; such a task might appeal to a few with great ambition, but it is simply too time-consuming and labor-intensive for most. Rather, those who engage in news-spreading practices seem to be driven largely by their individual political passions, passing on only those stories that address the issues they care about most or in which they are most personally invested. The result may indeed be a selective lens on the news of the day, but also one that can be highly engaged and deeply felt in an environment where political information is often disparaged as dry or boring. Furthermore, we have noted how these practices also promise to amplify a range of alternative and marginalized voices and decenter the agenda-setting power of established media elites. Again, we can glimpse how a dynamic of affective identification holds the promise of invigorating citizen-level participation in public issues at the same time as it clashes with certain idealistic notions of democratic life—in this case, a rational, deliberative public sphere in which a consensus of truth can be established.

In addition, we also saw in Chapter 5 how for-profit news agencies work to shape peer-to-peer flows of online political information in their own economic interests, and how this presents a further set of challenges for citizens who seek to impact the agenda-setting process and advance their political priorities in a substantive fashion. As digital news content is increasingly fashioned for viral transmission to maximize clicks and ad revenue, patterns of soft news that have long characterized commercial journalism may become even more pronounced. Since viral shares on social media appear to be largely driven by intense emotional responses,[26] for-profit news agencies are incentivized to cover political issues through the lens of dramatic and personalized narratives that risk trading complexity and depth for provocative and captivating surfaces. In other words, we find yet another manifestation of the light politics paradox, in which some of the substantive political content of partisan and advocacy journalism is sacrificed for the viral appeal of dramatic spectacle. As numerous scholars argue, emotionally resonant news narratives that borrow from widely accessible popular culture genres and formats can play an important role in boosting citizen engagement in the political sphere and bridging the gap between distant political institutions and everyday life.[27] This possibility may serve as a compelling defense for the production and dissemination of softer forms of online political news, yet at the same time, we can also recognize how the packaging of issues for popular circulation demands a degree of reduction and simplification. Whether the light packaging is ever removed to reveal the weightier material lying inside thus becomes a key question for those who wish to shift the public conversation in ways that support meaningful structural change.

Perched on a Baudrillardian Cliff?

The light politics concept can thus help to elucidate a fundamental challenge that appears to face all participatory political promotion efforts more broadly: retaining connections between the symbolic and the material, between cultural texts and political and social structures. Decades ago, the preeminent philosopher of postmodernity, Jean Baudrillard, observed the growing shift toward a fully media-centered world of images and signs and concluded—pessimistically or gleefully, depending on one's interpretation—that this would constitute "the implosion of the social."[28] In Baudrillard's theory of the four successive phases of the image, images begin as accurate reflections, or representations, of a tangible, material social reality, but eventually become detached from whatever they are intended to signify and enter a realm of pure simulation. In a postmodern, media-drenched age of what Baudrillard calls hyperreality, the fate of the image is to ultimately retain "no relation to any reality whatsoever."[29]

When applied to the domain of political media and marketing, such a theory points to some chilling conclusions. As the world of politics becomes increasingly inscribed within mediated circuits of symbolic representation, Baudrillardian theory suggests that these images and signs will ultimately lose any connection they once may have had to material politics—that is, the on-the-ground distribution of resources—and become pure simulations of the political. This is perhaps the most extreme version of the political trivialization critique discussed previously: here, the embedding of complex public issues within more attractive symbolic packaging threatens not just to dumb them down or flatten them, but to eradicate them completely. As popular participatory media practices like forwarding hashtags and memes help make political communication lighter— more entertaining, more accessible, more casual, more mobile, etc.—the risk is that political discourse will leave its material grounding entirely and float away into an ether of hyperreality. In other words, these practices seemingly threaten to push political communication further toward the precipice of what we might call the Baudrillardian cliff. At its tipping point, images and signs that ostensibly reference the structural political world shed these ties and become free-floating entities that refer only to themselves. Although they may serve as sources of great pleasure, passion, and identification as they are circulated freely and widely in popular cultural spaces, they may lose the power to even address—let alone influence or change—the underlying structures of society. Falling off (or, to continue the light politics metaphor, floating up and away from) the Baudrillardian cliff would thus present the ultimate doomsday scenario for citizens who wish to use participatory media platforms to substantively reshape the world in their interests.

Is this the ultimate fate of the citizen marketer approach—the endpoint of a decades-long trajectory in which citizen participation in political discourse is increasingly squeezed into the confines of catchy T-shirt and bumper sticker slogans, whimsical Internet memes, trendy profile-picture and hashtag campaigns, and virally pitched soft news packets? As critics of postmodern theory's most radical extremes would respond, not necessarily. One particularly compelling alternative framework is offered by André Jansson, whose theory of image culture recognizes the forces of culturalization, mediatization, and simulation that figures like Baudrillard draw attention to yet categorically rejects what he calls "the postmodern hypothesis of cultural implosion."[30] In its place, Jansson asserts that meaning within an image culture is maintained in "the hermeneutic activities of social actors,"[31] which includes the agency of individuals and groups to actively decode or interpret the meaning of the images and signs that they consume, as well as their capacity to re-encode meaning through creative practices of cultural expression and circulation.

Here, Jansson extends Stuart Hall's cultural studies model of the active media audience[32] to the domain of consumer culture, which he argues has become completely inseparable from media culture. Extending this line of thought further, we can apply Jansson's hermeneutic perspective to the promotional practices of grassroots intermediaries—which, as discussed in Chapter 1, grow directly from consumer culture—including those that involve political forms of brand evangelism. Such a perspective would emphasize that political meaning is not inherently present or absent in the symbolic media content that circulates across citizen networks, but rather is continually reconstituted as this content is interpreted, modified, and shared by groups of social actors. Just as selective forwarding itself involves active processes of expanding and building upon existing meanings through curatorial agency (as discussed in Chapter 1), the political significance of any peer-to-peer political marketing effort hinges on the active interpretive work of those who encounter and help spread it.

The hermeneutics-focused perspective thus serves as a comforting rejoinder to overly apocalyptic postmodern visions of the implosion of meaning and the death of the social. However, it does not simply ignore or cast aside the profound tensions that are inherent in light politics. Rather, it suggests how the onus of

on three areas of concern that are crucial for situating citizen marketing in a broader structural context: issues of unequal technological access and skills that differentially impact citizen participation in online political persuasion; the role of powerful political and economic elites in the shaping of peer-to-peer message flows; and the overarching ideologies and logics of neoliberal market capitalism that seemingly fuel, and are potentially reinforced by, the conflation of democratic citizenship and marketing logics.

Who exactly comprises the emergent group of media-based symbolic participants, or the critical periphery, in political movements, causes, and campaigns? Are there segments of the public that are systematically left out? Although it is difficult to say with certainty, we can assume that numerous structural factors impact and moderate participation in citizen marketing practices. The respondents interviewed for this book, although they are not scientifically representative of the broader population of all those who engage in citizen marketing activities, suggest some tentative clues as to the characteristics of this population: by and large, they are highly educated (often with a postgraduate degree), consume a high level of political news on a regular basis, and note a strong interest in political topics more generally. The picture painted is thus one of a narrow group of so-called political junkies who may hold relatively privileged positions within society.

However, it would be a mistake to assume that citizen marketing practices are, by definition, exclusionary or elitist. Although there may be systematic biases in terms of who ultimately engages in such practices, the low cost and relative accessibility of the participatory media platforms in question, as well as the broad appeal of the largely popular culture-oriented content that circulates across them, suggest that barriers to entry are significantly porous in nature. As Karpf emphasizes, the subset of the population that uses social media to express their political views may only be a small fraction of the citizenry, but the "activated public opinion" they produce is "versatile and more publically accessible" because of the open-ended nature of networked digital technologies.[35] Indeed, the embrace of online citizen marketing practices by a range of traditionally marginalized and outside-the-mainstream constituencies, from members of the LGBT community to members of protest movements structured around issues such as income inequality and racialized police violence, suggest that the phenomena in question do not simply represent the kind of "politics as usual" predicted by normalization theory—that is, that political elites will ultimately seize control of Internet technology to reinforce their own power.[36]

Still, it is important to acknowledge how broader power relations, such as those involving the digital divide, nonetheless impact the phenomenon of citizen marketing. Although this book has been concerned with offline as well as online platforms of expression, including the use of material culture artifacts that can be as simple and low cost as a handwritten sign, the fact that so much symbolic political action now takes places in networked digital spaces means that the less than fully democratic nature of the Internet cannot be ignored when assessing its democratizing potential. According to 2015 data from Pew Research, Internet access and use in the United States remains uneven, with significant gaps related to differences in household income, education, race, age, language (English versus Spanish speaking), and community type (rural versus urban).[37] Quite simply, groups that are less likely to be accessing the Internet in the first place are also less likely to be well represented in scenes of online political opinion expression.

In addition to gaps in access, the discourse of the digital divide also draws attention to structural gaps in the skills needed to use the Internet effectively, such as the ability to gather information and use online tools for strategic purposes.[38] Although the operational skills required to perform symbolic actions on social media such as posting profile pictures, circulating hashtags, and linking to articles may themselves be relatively easy to master, the ability to use these techniques effectively to advance one's political interests may be unevenly distributed among Internet users, since they draw on informational and strategic skills that are correlated with differences in education (among other interrelated variables).[39] Therefore, for citizen marketing in the digital age to live up to its potential of democratizing the field of mediated political persuasion that has been traditionally dominated by elites, these divides in access and skills must be addressed.

In his discussion of how to reform the Internet to better serve the needs of democracy, Robert McChesney outlines a detailed set of policy proposals that will be crucial for remedying such imbalances in the future, such as "broadband availability to all for free as a basic right" and "comprehensive media literacy education in schools to give people a critical understanding of digital communication."[40] Similarly, Howard Rheingold stresses that youth media literacy education is necessary for helping young people develop the ability to "communicate in their public voices about issues they care about" and thus participate in projects of advocacy and influence that represent their social and political interests.[41] As we recognize the interpretative work required to make citizen marketing more meaningful and more tethered to political realities, we must also recognize that the skills needed to engage in this work may be unevenly distributed because of existing social inequalities. Thus, supporting media literacy (as well as increased technological access) among disadvantaged groups can help empower them to take full advantage of these practices and tactics for their own purposes. For those who are invested in the project of democratizing the Internet, including its uses for symbolic political action, advancing these policy initiatives will be imperative moving forward.

When considering issues of inequality in citizen marketing's digital spaces, it is also necessary to address the fact that the commercial platforms commonly used for political media-spreading activities, such as Facebook and Twitter, are hardly neutral communicative spaces. In recent years, there has been increasing focus on the structural constraints that these platforms place on peer-to-peer expression, such as the practice of algorithmic news feed filtering discussed in Chapters 4 and 5.[42] If the reach of citizens' posts and links is systematically limited to connections with whom they interact the most and who likely share similar interests and views—or if their expressions are pushed out of others' feeds altogether because of their lower status and position in the network—then this could seriously impact their capacity to contribute to broader media flows in ways that serve their political interests. Furthermore, growing concerns exist regarding politically inflected censorship practices on commercial social media platforms such as Facebook, ranging from the banning of mothers' breastfeeding images to the suspension of pages organized around controversial protest movements,[43] and this also pushes back against the notion that these platforms are fully open and democratic venues for citizen expression and political promotion. In addition, there is also the matter of state censorship of online communication in many countries around the world that are ruled by authoritarian regimes, which includes both the targeted deleting of political posts[44] and the disconnection of Internet service more generally during contentious periods.[45] Together, these issues highlight how a well-rounded critical literacy of citizen marketing— not to mention of digital communication more broadly—must include an acute awareness of, and active engagement with, the structural barriers and hurdles posed by commercial digital intermediaries and the governments that regulate them. Although the Internet may represent a considerable advance over earlier media technologies by expanding pathways for popular participation in a range of expressive actions, there is a long way to go before it lives up to its promise of enabling a truly democratized system of political communication.

In addition to these issues of inequality related to the technological and legal structures that govern digital political expression, this book has also stressed the multiple ways in which the power relations of economic and institutional elites bear upon the practice of citizen marketing, particularly in terms of top-down efforts to steer, manage, and control peer-to-peer flows of political messages. Although we can recognize elements of grassroots citizen empowerment in the shift toward a more participatory model of political promotion and persuasion, we have also seen how powerful institutions and organizations are attempting to shape this activity to suit their needs. In Chapter 3, we noted that large and well-funded nonprofit advocacy organizations like the HRC have adopted sophisticated viral marketing techniques to help promote their agendas and unify social movements under their branded banners, as well as how this may threaten to crowd out space for the diversity of expression within these movements. In Chapter 4, we examined the efforts of election campaign organizations to similarly harness the force of citizen marketing via top-down digital marketing tactics designed to transform supporters into surrogate carriers for campaign messages, as well as broader media image-crafting practices that encourage fanlike evangelizing at a more grassroots level; in both cases, these efforts have the potential to direct citizen marketing activity in ways that can enhance elite institutional power. Finally, in Chapter 5, we considered the economic incentives of for-profit journalism agencies to exploit the politically charged sharing of online news articles to maximize advertising revenue, resulting in "click-bait" coverage and soft news values that may conflict with the politically-minded goals of participatory agenda setting. In each case, we recognize that the promotional labor of citizen marketing, which is ostensibly aimed at publicizing and amplifying one's political opinions, ideas, and priorities, simultaneously creates promotional value for a range of elite institutional actors whose content is being circulated and endorsed. At the same time, it appears that many citizens are eager to take on a deliberate brand evangelism role in these contexts, so long as their political interests are in alignment with those of the respective entity in question.

For those who are drawn to the citizen marketer approach to political action, assessing these potential alignments—and potential misalignments—is crucial. Thus, a key aspect of developing a critical literacy regarding these practices involves heightening one's awareness of the *sources* of the content he or she promotes through selective forwarding practices, as well as the institutional interests represented therein. This may seem obvious in some contexts, such as sharing a link to an electoral candidate's campaign ad, but less so in others. For instance, during my interviews with citizens who posted the red equal sign profile picture on Facebook, I was struck that several respondents were unaware that the campaign was designed and orchestrated by the HRC, which is particularly significant considering its controversial status in the broader LGBT movement (as discussed in Chapter 3). As the speed of political message flows grows ever faster in the age of social media and its attending cultural norms of viral trendiness and memetic replication, there may be less time for citizens to fully contemplate what exactly is being shared and who ultimately stands to benefit. Again, this brings us back to the importance of adopting a reflexive approach to citizen marketing—in this case, thinking through the repercussions of one's promotional labor in the political media landscape and its embeddedness in larger institutional power relations.

Furthermore, the very notion of promotional labor as a political act may be anathema to certain portions of the citizenry that are ideologically opposed to the neoliberal market relations and logics that have fueled the rise of promotional culture in Western capitalist societies more broadly, including in the political realm. For conservatives, whose worldview tends to affirm the ideology of market capitalism as morally sound and socially beneficial, the prospect of conflating political activity and marketing activity does not appear to pose much of a dilemma. Indeed, among the conservative-identified citizens interviewed for this book, the framework of marketing was often embraced without reservation when explaining the goals of their political media-spreading activities. However, among the more left-wing and liberal-identified respondents, there was a divide between those who were mostly comfortable with the thought of behaving in a marketing-like capacity to spread their political opinion and influence and those who were either ambivalent or outright averse to such a notion. This outcome is likely a sign of the broader tension regarding normative perceptions of marketing in left political circles, which boils down to the question of whether its practices and logics necessarily affirm neoliberal market relations or can alternately be used to challenge them. For those who stand in opposition to the ideology of market capitalism and who struggle to resist it, this is perhaps the most treacherous terrain that a critical literacy of the citizen marketer approach can potentially help to navigate.

In the words of Micah White, the co-founder of the Occupy Wall Street movement along with his *Adbusters* colleague Kalle Lasn,[46] "a battle is raging for the soul of activism. It is a struggle between digital activists, who have adopted the logic of the marketplace, and those organizers who vehemently oppose the marketization of social change."[47] White specifically singles out the progressive website MoveOn for its use of sophisticated digital metrics to track and target its audiences in professional marketing fashion, a point that brings to mind the earlier-noted critiques of electoral political marketing as positioning voters in an atomized, consumer-like role.[48] He goes on to argue that "any activism that uncritically accepts the marketization of social change must be rejected." White's complaint that left activism has embraced marketing logics to its detriment may seem ironic on the surface, given his *Adbusters* outfit's efforts to co-opt marketing's forms and languages for radical political ends. As discussed in Chapter 1, the *Adbusters* culture-jamming model seeks to appropriate and subvert marketing discourses and turn them against neoliberal capitalist ideologies. Crucially, such an approach fundamentally accepts—and indeed operates from—the premise of a fully postmodern political reality in which mediatized promotional culture is all-encompassing and inescapable. As I have argued, this is also the core underlying assumption of the citizen marketer approach to political action more broadly, regardless of the ideological positions of those who adopt it. If such a premise is in fact the reality that all citizens and constituencies now face, then the call for marketing logics to be wholly cast aside within the realm of political action would appear to be untenable.

However, a key distinction can be made. White's critique is concerned with large and well-funded political advocacy organizations that conduct themselves like commercial marketers through extensive research and metrics and the technocratic tracking of user clicks and interactions—the very sort of elite institutional activity discussed previously. Such a position dovetails with Heather Savigny's broader criticism of what she calls the antidemocratic tendencies of electoral political marketing, which focuses on a business-derived managerial model that serves short-term institutional interests rather than the long-term interests of the larger polity, and makes questionable claims of being democratic by merely pointing to campaign adjustment processes based on public opinion research.[49] Similarly, for White, the critique of the marketization of activism appears to be aimed at the hierarchical relationship by which technocratic organizational leaders act as the marketers and citizens act as consumers whose agency is restricted to clicking their approval of promotional appeals and providing limited forms of "customer feedback."

By contrast, although White's and Lasn's outfit, *Adbusters*, launched catchy marketing-like hashtag slogans and graphics to promote the first Occupy Wall Street protests and spur citizens to mobilize in the streets of New York City, the resulting movement was notably grassroots, decentralized, and unmanaged.[50] That is to say, White and his colleagues appropriated marketing techniques to grab the public's attention and promote an influential idea, but then avoided acting as professional political marketers in the sense of top-down management, research, and metrics. Likewise, in *Culture Jam*, Lasn calls on citizens to become what he calls meme warriors in the grassroots spread of subversive political messages,[51] which suggests a more democratized—or at least more populist—reworking of marketing logics in the service of radical political activism. Instead of asking powerful organizations to manage consumer-like citizens through forms of controlled interactivity, Lasn urges grassroots activists to claim their own marketing agency—to become empowered players in the circulation of persuasive political messages. These activities may certainly be used in the service of advancing one's own self-interest, much in the way that critics of neoliberal market ideologies and their effects on citizens might predict. However, they may also be utilized collectively—or, more likely, connectively—by groups that have been historically marginalized and underrepresented in the traditional marketplace of ideas dominated by elites.

Is it therefore possible to adopt the languages and styles of marketing for political persuasion and advocacy, but not its corporate-like administrative structure and attending values of instrumental economism? Critics of the culture-jamming approach contend that opening the Pandora's box of marketing perennially risks incorporation by the economic status quo.[52] However, this high-wire balancing act appears to be crucial for those who locate neoliberal market relations as a key component of political struggle and yet still wish to use the persuasive power of media-based symbolic action to their political advantage. To be certain, the meme warrior alternative to the controlled interactivity of institutional political marketing does not entirely escape the ideological pull of marketing logics and promotional culture. Yet rather than simply adhere to these logics and reproduce their power hierarchies—the *uncritical* acceptance of marketing in White's formulation—it attempts to subvert and re-invent them as a kind of *critical* acceptance. Like many of the thorny dilemmas explored in this chapter, it would appear that such nuances can only be successfully navigated through concerted reflection and self-awareness of practice.

Conclusion: Meme Warriors on the Media Persuasion Battlefield

We can thus suggest that the meme warrior conceptualization of citizen marketing holds the promise of a more *maximalist* form of participation in political promotion and persuasion (to borrow from the vocabulary of Nico Carpentier),[53] in which citizens take on a co-deciding role regarding what messages get circulated in the political media landscape. As this book has argued, such potential for citizen empowerment rests not only in the creative agency of amateur media production, but also in the curatorial agency of shaping and (re)directing persuasive media flows through practices of selective forwarding. Regarding the normative theories of the public sphere treated in Carpentier's analysis of media and participation, this ultimately resembles less the classic marketplace of ideas of liberal political theory, which implies a limited set of choices handed down by institutional elites, than the more radical-democratic models that position the public sphere as a "public arena of contestation."[54]

The image of the meme warrior eloquently captures this notion of symbolic contestation at the level of the ordinary citizen. Furthermore, it signals the potential instrumental power of peer-to-peer media circulation as the aggregate and networked spreading of influential ideas, which has too often been overlooked in analyses that either categorically dismiss symbolic action as ineffectual slacktivism or evaluate it only in terms of its capacity to lead to other forms of political participation down the line. However, although this suggests a degree of citizen power that has traditionally been missing from the top-down machinery of political marketing, the practices explored in this book are not immune from elite management and control. Indeed, such a point underlines the importance of developing a critical literacy of citizen marketing that can uncover the power relations embedded in hybrid political media flows and gauge one's own position and participation within them.

In addition, the image of the meme warrior reminds us of how broadening the circle of participation in mediated political persuasion reinforces, and likely intensifies, an agonistic vision of democracy in which civic discourse is marked by symbolic conflict and confrontation rather than by deliberation and consensus building. As the firsthand accounts in this book suggest, there do appear to be cases in which bold displays of political opinion and partisan identification in casual cultural spaces open the door to two-way civic exchanges and mutual learning and connection, suggesting a certain degree of fluidity between the citizen marketer approach and the models of deliberative democracy and civic cultures. In other cases, however, these acts of declarative side taking and statement making may exacerbate the dynamics of political polarization and retrenchment that have divided the polity along tribal-like, culturally grounded lines. As contemporary forms of symbolic political action increasingly compel citizens to draw on the persuasive force of their own identities to stand behind and authenticate political messages—whether through the self-expressive spaces of their social media profiles and avatars or through the signifying spaces afforded by their corporeal bodies and material possessions—a likely consequence is the deepening of partisan identifications and the conflation of cultural and political identity. However, although many democratic idealists may recoil at the notion of a factionalist, even sectarian, citizenry pitted against each another in acts of media and information warfare, others may welcome the entrance of historically marginalized constituencies into this battle over the minds as a result of the opportunities introduced by participatory media channels.

Finally, the invocation of the meme in meme warrior brings us back to the dilemmas and trade-offs of light politics, which appear to loom over all political constituencies that seek to appropriate modern media marketing techniques as a means of spreading political opinion and influence. Although the term meme has been used to refer to any idea that is replicated across a culture,[55] its common contemporary association with bite-size and fast-traveling packets of symbolic digital content, often humorous and irreverent in nature, effectively encapsulates both the lightness of much political expression that is circulated peer-to-peer and its lightness of rapid movement across cultural spaces and social networks. Although the light politics of meme culture has the potential to serve democracy well by carving out a space for emotion as well as reason in the political sphere and expanding accessibility and popular engagement, it also risks teetering off the Baudrillardian cliff into empty simulation if the balance between style and substance is thrown askew. As the legacy of propaganda, the forebearer of political marketing, reminds us, the aestheticization of politics can all too easily serve as a breeding ground not only for trivialization, but also for mass manipulation. At a time of expanding participation in persuasive political communication, such concerns are applicable not only to the elite institutional forces that continually seek to shape political media flows, but also to all who step onto the battlefield of meme warfare. Indeed, Nicholas O'Shaughnessy argues that "classical propaganda was something produced by powerful factions and forces or governments or parties. The rise of the Internet has changed that balance, with YouTube or social networking allowing private citizens to sponsor their own propaganda campaigns."[56] Although it may be hyperbolic to suggest that the citizen marketer is ultimately one and the same as the citizen propagandist, the latter's negative connotations and historical baggage underscore the profound tensions and challenges that await those who enter the fold of mediated, packaged political persuasion.

As I have emphasized in this concluding chapter, fostering a more reflexive and more thoughtful approach to this set of practices is likely our best hope for avoiding its greatest potential hazards. Citizen marketing and its attending dynamic of light politics is not a trend that can be simply dismissed or pushed aside because of unsavory or worrisome aspects. Rather, it represents an increasingly central logic of what it means to be an engaged and active citizen in an era of image culture, postmodern politics, and peer-to-peer connectivity. Citizen marketing has in fact a long and significant history in the field of democratic practice and is reaching new heights of popular participation as a result of key cultural and technological developments. For better or for worse, the citizen marketer approach is here to stay, and the most productive path moving forward is to try to make it more introspective, more adaptive, more critically engaged and aware of its embedded power relations, and perhaps, ultimately, more meaningful.

Methodological Appendix

The interview data presented in this book are culled from three separate research studies conducted from 2010 to 2013, each focusing on a different aspect of citizen-level participation in the spread of political media content. Results from each study have previously been published in various journal articles (as indicated below), although much of the data included in this book have yet to be published in any form. Since the in-depth interviews produced far more data than I was able to discuss in these publications, my approach for this book has been to revisit this rich data set and significantly expand on the earlier analyses. In this appendix, I detail the methodology for each of these studies and recount the broader investigative research process that led to their formation.

My research on what I term the citizen marketer approach began as an outgrowth of my dissertation work at the Annenberg School for Communication at the University of Pennsylvania. At the outset, my focus was on the specific practice of wearing T-shirts imprinted with political messages and slogans, and the research sought to uncover citizens' motivations for engaging in this practice. Entitled "Body Screen/Body Politic: The Uses of Political T-Shirts in the Digital Age," the project was oriented by the grounded theory approach,[1] which involves an inductive process of theory building through systematic analysis of qualitative empirical data. Rather than testing preformed hypotheses, I sought to explore questions of motivation through open-ended and exploratory conversations with interview respondents. Following grounded theory protocols, I systematically coded the interview data by theme and then identified broader conceptual categories to account for these codes. Through this process, the concept of the citizen marketer approach—that is, the motivation to promote one's political viewpoints to peers in a viral marketing–like fashion by participating in the dissemination of symbolic media artifacts—emerged organically from the data as a metacategory of analysis. This key finding formed one of several conceptual frameworks of the completed dissertation, and I decided to further pursue the development of this framework through additional empirical research projects. In particular, I was fascinated by how numerous interview respondents compared their persuasion-oriented motivations for publicly wearing political T-shirts to similar motivations when posting political media content in online venues, such as profile pictures on Facebook. This insight inspired the subsequent direction of my research for several years to come and shifted my focus to digital scenes of citizen-level media circulation that were addressed only intermittently in the dissertation study.

Before outlining the methodological details of the initial dissertation project and the two additional studies that followed, I will address the general research approach that guided all three projects. Each study was grounded in a "naturalistic, qualitative social research" orientation that is concerned with the ordinary settings in which people live and operate and sought to "convey to others, in rich and realistic detail, the experiences and perspectives of those being studied."[2] In each case, I was guided by the notion that "at the root of in-depth interviewing is an interest in understanding the experience of other people and the meaning they make of that experience."[3] Since my primary research interest involves questions of subjective personal motivation, I opted to use an open-ended, semistructured, in-depth respondent interview method, of which one of the primary goals is "to understand the interpretations that people attribute to their motivations to act."[4]

In other words, my goal was to ground the investigation in the respondents' own meanings and logics, rather than ascribe explanatory frameworks from my own vantage point. In doing so, I sought to uphold the ethical philosophy of the naturalistic qualitative research tradition, which involves treating the people with whom I am studying "as partners rather than . . . as objects of research."[5] Instead of attempting to unearth hidden or subconscious motivations for the respondents' practices, I committed to taking their subjective accounts seriously and using their own conceptualizations of practice as the backbone of my grounded theory building. Thus, rather than test hypotheses of motivation with specific probes, I focused on asking open-ended questions that gave respondents the space to articulate their own ideas about what they were doing with regard to their political media participation and why they were doing it. Following the technique of responsive interviewing,[6] the conversations moved in a unique direction for each interview and adapted to each respondent's individual reflections.

When analyzing the respondents' accounts, I made a deliberate choice to read them at face value, rather than against the grain. In other words, I accepted their interpretations of practice in a relatively straightforward fashion and faithfully rendered them in my analysis without searching for alternate or hidden meanings. Although it is possible that what the respondents thought they were doing was not what they were actually doing (i.e., they may have been unaware of or unable to fully articulate certain underlying or subconscious motivations for their practices), I opted to stay close to what they verbally articulated in the interviews rather than attempt to complicate their interpretations by introducing my own. Admittedly, this decision to read the interview data at face value effectively set aside the issue of the respondents' possible unspoken agendas for agreeing to be interviewed regarding their political media-spreading activities. I did not specifically focus on how their conversations with me as an academic researcher might have been *invested*—that is, how they may have had an interest in publicizing and justifying their specific political media practices and corresponding viewpoints (as well as the groups which they may have felt to be representing), thus shaping their responses according to what they imagined to be a desirable outcome of the exchange. Indeed, the respondents may have seen the interview itself as an opportunity to champion their positions and causes, as well as to defend their particular media-spreading activities as politically valuable and useful. My point here is not to call into question the veracity of their responses, but to stress that their respective political investments may have influenced how they framed their experiences in certain ways.

This set of choices was guided by my research questions that center on subjective motivational thought processes, as well as my personal philosophical orientation as a naturalistic qualitative researcher with the primary interest of conveying to others the experiences and perspectives of those whom I study.[7] However, it is important to acknowledge that other approaches to analyzing interview data may have led to different outcomes. Specifically, taking a more skeptical approach to the respondents' accounts may have complicated some of the findings regarding the underlying reasons for spreading political media content to peers. For instance, it is possible that some respondents had an investment in convincing me of the political effectiveness of their particular political media-spreading practices (e.g., for persuading others or, alternately, for sparking deliberative discourse), and this positive and optimistic emphasis may have colored the resulting analysis to a certain extent. Thus, my choice to read the respondents' accounts at face value can be recognized as a limitation of each of the three studies, albeit one which has its strengths as well, particularly with regard to conveying the respondents' subjective meanings and conceptualizations of practice in naturalistic terms.

Study 1: Wearing Political T-shirts

The first of the three studies involved my dissertation research that focused on the act of wearing political T-shirts (defined as T-shirts imprinted with politically oriented content in the broadest possible sense). To recruit volunteers for in-depth respondent interviews, I employed the strategy of theoretical sampling, which refers to "choosing those whose testimony seems most likely to develop and test emerging ideas."[8] Following the comparative method of analysis recommended in the grounded theory approach, I used "the systematic choice and study of several comparison groups"[9] to develop broader conceptual categories that cut across individual cases. In other words, rather than attempting to provide a representative sample of all political T-shirt wearers located in the United States, I opted to strategically choose comparison groups that would help to build new theory in accordance with my research objectives. Specifically, I targeted recruitment at four broadly defined comparison groups that were deliberately chosen to increase variation within the data: Democrat supportive, Republican supportive, LGBT issues supportive, and environmental issues supportive. These choices reflected my interest in including perspectives from both liberals and conservatives, as well as from people who are engaged at the institutional level of politics (i.e., elections) as well as the noninstitutional level (i.e., social movements). In fact, substantial overlap existed across these groups, particularly among those who wore T-shirts supporting Democratic politicians such as Barack Obama and those who wore T-shirts supporting the left-leaning social issues of environmentalism and LGBT rights. Since the interviews were open ended and not bound by any particular political subject matter, the comparison groups served as slices, or entry points, of the broader populations of left-wing and (to a lesser extent) right-wing political T-shirt wearers.

Because of the logistical constraints of finding appropriate volunteers who qualified for the comparison groups, the recruitment procedures relied on a combination of convenience and snowball sampling techniques that primarily targeted college students in the Philadelphia area who were on the email contact lists of relevant political organizations. Recruitment messages were emailed to campus organizations at seven different universities in the region (e.g., College Democrats, College Republicans, LGBT student groups, environmentalist clubs), and volunteers were encouraged to share the recruitment messages with others who may have qualified for the study and were interested. In addition, some volunteers were recruited by handing out fliers at relevant events and gatherings in the Philadelphia area (e.g., a Barack Obama rally, a gay pride event), which was intended to diversify the respondent pool beyond college students and those with organizational affiliations. Although eight respondents were recruited in this fashion, some shared the same characteristics of being enrolled in college and being members of campus political organizations.

The resulting sample of 54 respondents has multiple limitations to generalizability and is intended to serve as a partial window into an indeterminate population of U.S. political T-shirt wearers that is nonetheless valuable for theory building. Of these respondents, 44 were currently enrolled as college undergraduates in the Philadelphia area, which indicates significant limitations with regard to age, level of education, geographical location, and likely socioeconomic status. In addition, the majority of respondents reported a strong level of interest in political topics, as well as prior experience participating in political organizations (typically campus based). Since there are no representative statistics available regarding the broader population of political T-shirt wearers in the United States, it is unclear to what extent these recruitment outcomes are related to broader trends among this group. For instance, it is possible that college-age young people are more likely to participate in this activity than their older counterparts, as are those who are organizationally affiliated versus those who are not. Yet regardless of the sample's true level of representativeness, the 54 in-depth interviews yielded a variety of contrasting responses that suggested a diverse range of perspectives among those who participate in this practice. Indeed, the notable differences that appeared between some respondents who readily embraced a marketing-like framework of peer persuasion on their own accord and others who were ambivalent or resistant to such a notion inspired the key conceptual schema for the resulting analysis (as well as the approach of this book as a whole).

In terms of including gender and racial diversity to increase the variation of perspectives, the sampling process was somewhat successful. Overall, 31 respondents were female and 33 were male, providing nearly even representation for each sex. Forty-one respondents identified as white, 9 as African American, and 4 as Asian American. Regrettably, the sample did not include any respondents who identified as Hispanic or Latino (another important limitation of this study). The respondents' ages ranged from 18 to 31, with a median age of 20, which again signals the youth-oriented focus of the resulting data. With regard to the comparison groups, the sample was fairly well distributed: 16 respondents were included in the Democrat-supportive group, 12 in the Republican-supportive group, 13 in the LGBT issues-supportive group, and 13 in the environmental issues–supportive group.

The semistructured interviews were conducted in person on the campus of the University of Pennsylvania during a three-month period in late 2010 and early 2011. Each interview lasted approximately 30 to 90 minutes, depending on the direction of the conversation. Respondents provided informed consent prior to being interviewed, which included giving permission to use their data for future research projects. All identifying information was made confidential in the transcripts of interview recordings, and respondents were given pseudonyms to protect their privacy. Portions of the data were published in two separate journal articles in *The International Journal of Communication*[10] and *Popular Communication*.[11]

In addition to questions regarding basic demographic background information, the schedule of interview questions was as follows:

Please describe the political T-shirt(s) you personally own.

Where and when did you acquire the T-shirt(s)? Please describe the circumstances in which they were purchased or otherwise obtained.

Do you own other similar political T-shirts? Please describe them.

Why did you choose to acquire this particular T-shirt(s)?

Why did you choose to acquire a T-shirt for this specific cause/candidate?

What do you like about the design of your T-shirt(s)? Why did you choose it from among others?

What does your T-shirt(s)' design mean for you?

In which places and in which social situations have you worn your T-shirt(s)?

How often have you worn your T-shirt(s) in these different places and situations?

Did your wearing of the T-shirt(s) coincide with any relevant political events?

Are there any other patterns of wearing you can identify?

Are there preferred circumstances for wearing your T-shirt(s), and why?

Are there places or situations where you wouldn't wear your T-shirt(s), and why?

Do you think that your T-shirt(s) communicates anything to other people who see it? If so, what?

If you see your T-shirt(s) as communicating to other people, who do you see it as addressing?

How have other people responded to your T-shirt(s)? Describe any stories you recall involving people's reactions, positive or negative.

Has your T-shirt(s) ever gotten an unpredictable or surprising reaction?

What do you see as your motivations for wearing your T-shirt(s)?

Did these motivations change over time? Were they different for specific places and/or situations?

Please briefly describe your major activities which you see as relating to your political identity. This may include working/volunteering for political organizations, attending rallies, meetings, conventions, etc.

Please briefly describe any other kinds of relevant political communication you engage in beyond wearing your T-shirt(s). This may include online communication (blogging, social network messaging, emailing, etc.) or offline communication (displaying bumper stickers, lawn signs, and/or posters, conversing with friends, family, and/or neighbors, etc.)

How do you see your T-shirt(s) in relation to these other kinds of relevant political activities and/or communications which you have been involved with?

Study 2: Sharing Election-Themed Content on Twitter

The second respondent interview study included in this book was conducted in the context of the 2012 U.S. presidential election cycle and was intended to explore the motivations behind sharing election-themed media content on the popular social media platform Twitter. Although the citizen marketer framework that had emerged from prior research served as a sensitizing concept, this study continued the exploratory and naturalistic approach that I had established in the previous study, in keeping with the rationale outlined above. The notion of marketing-like peer persuasion was not introduced upfront as a prompt or probe, and respondents were given the opportunity to discuss their motivations (as well as their reflections of practice more broadly) in an open-ended conversational format. However, if and when such a notion arose organically in the respondents' remarks, follow-up questions were used to further probe the contours of this framework, as well as other contrasting frameworks that were articulated in the course of the interviews.

Although the goal of the study was to initiate a wide-ranging set of conversations about the practice of spreading election-themed media on Twitter, I chose to focus on a particular piece of content as an entry point, or gateway, for these conversations. This had the benefit of providing practical direction for recruitment and also established a common baseline for comparison across the different interviews. Specifically, I targeted recruitment materials at Twitter users who had shared a link to a specific YouTube video, entitled *Will the Real Mitt Romney Please Stand Up?* This video was selected for several reasons: first, it was one of the most popular viral videos at that point in the 2012 presidential election cycle, garnering more than 5 million views; second, following theoretical sampling principles, I identified it as an ideal case study to explore research questions relating to personal motivation because the precise purpose of both the video and its viral circulation was not readily apparent. Although the video appeared to contain an anti–Mitt Romney bent and could therefore be perceived as a form of negative campaign advertising, it was produced without the involvement of any campaign organization or political group (rather, it was made by the Australian amateur YouTuber Hugh Atkin) and featured satirical elements that seemingly blurred the line between popular culture entertainment and political statement making. The video was thus selected as an archetypical example of the popular viral media surrounding the 2012 presidential election, and the interviews sought to unpack how those who shared it understood the meaning of this kind of election-themed content as well as their reasons for selectively forwarding it to their peers on Twitter.

After making the decision to focus on one specific act of content sharing as the starting point for the study, the recruitment process was straightforward: the YouTube link was used as a search term on Twitter, and all users who had shared the link within the previous seven days appeared in the search results and were sent a brief recruitment message. All U.S. Twitter users over the age of 18 who had posted the video link were eligible to participate, resulting in a nonrandom sample that was diverse in some respects yet homogenous in others. Of the 25 respondents in the sample, 14 were female and 11 were male, with ages ranging from 18 to 70 (with a mean of 34.7). The respondents resided in 20 different U.S. states that represented all geographic regions of the country. The racial diversity of the sample was more limited, with 21 self-identifying as white, 2 as Hispanic, 1 as black, and 1 as multiracial. Reflecting the anti-Romney character of the YouTube video, the respondents largely self-identified as politically liberal, although the sample did include a Republican supporter of Romney's rival Ron Paul. In addition, the respondents uniformly reported a high level of education (some college and above), and 16 reported prior experience participating in political campaigns, organizations, and/or activist groups. Thus, the sample was skewed to a largely white, highly educated, liberal-leaning, and politically involved group of adult U.S. Twitter users, which suggests clear limitations to generalizability. However, such a small-scale sample is intended not to be representative, but to yield rich and in-depth qualitative accounts—however partial and incomplete—that aid in theory development.

The semistructured interviews were conducted over the phone during a two-month period in late 2012, and each lasted approximately 30 to 60 minutes. Respondents provided informed consent prior to being interviewed and later granted permission to use their data for future research projects. Transcripts of interview recordings removed all identifying information, and respondents were given pseudonyms to protect their privacy. Like the earlier political T-shirt study, the qualitative data analysis followed the grounded theory process: the data were first read over multiple times and systematically coded by theme, and these themes were then used to formulate broader conceptual categories that were modified and reworked until they represented the full scope of the data. Portions of the data were published in an article in *Convergence: The International Journal of Research into New Media Technologies*.[12]

The interview schedule consisted of the following questions (in addition to questions asking for basic demographic background information):

Tell me the story of posting this particular video on Twitter.
Have you posted videos like this before online? If so, briefly describe your
    previous experiences in posting these types of videos.
How did you choose this particular video to post?

How did you initially find the video?

What is the meaning of the video in your view?

If you added a comment while posting the video, why did you do so? What did you write, and how did you decide to write it? If you did not add a comment, why did you not?

What audience were you trying to reach while posting this video?

Who do you think actually saw the video from your Twitter feed?

Did your Twitter post receive any responses? If so, what was the nature of these responses?

What do you think were your goals in posting this video to Twitter?

Do you think these goals were met? If so, why? If not, why not?

In your opinion, what do you see as the general value of posting things like this to Twitter?

Please tell me about your political activities more broadly. This may include participating in political organizations, attending political events and rallies, and having political conversations with others in person or online.

Generally speaking, how would you describe the strength of your participation in politics?

How do you see your posting the video in relationship to the political activities you just described? Do you think they are connected or separate?

Study 3: Posting Gay Rights–Themed Profile Pictures on Facebook

The third respondent interview study included in this book continued my research trajectory of exploring digital contexts of political media spreading, although the focus this time was on the circulation of social movement–themed content. Specifically, my goal was to examine expressive digital practices related to the issue of gay and lesbian rights, which would expand on my earlier research on the display of LGBT-supportive messages via T-shirts. I chose to use the 2013 red equal sign Facebook profile picture campaign (discussed in detail in Chapter 3) as a case study, largely because it was one of the most high-profile social media events of the year, with millions of estimated participants and an accompanying wave of mass media coverage. In addition, I was particularly interested in investigating the phenomenon of profile picture–changing actions, since, like wearing slogan T-shirts, they appear to be grounded in the expression of one's personal identity (indeed, several respondents who were in the political T-shirt study drew this connection with political profile pictures in their comments). However, although previous research findings again provided key sensitizing concepts, the study followed its predecessors in adopting an open-ended and exploratory approach. As in the earlier studies, the framework of marketing-like peer persuasion was not introduced to respondents and was only addressed in follow-up questioning if respondents brought it up on their own accord without prompting. Other motivational frameworks were explored as well, in keeping with the principle of allowing respondents to share their own experiences and perspectives in a naturalistic context.

The recruitment process for the study presented a challenge, since unlike Twitter posts, Facebook posts are not typically made public to all users. In addition, the site does not allow for searches by profile picture. To find appropriate research volunteers, I tailored the recruitment to the social setting of Facebook and tapped into my own social network on the platform. Specifically, I posted a recruitment message to my own Facebook connections (I had approximately 500 at the time) that included a request to repost the message to their own respective connections. Speaking reflexively, this choice reflects my own social position as a U.S. adult Facebook user who is "friends" with many politically liberal, LGBT, and LGBT-supportive users. To minimize potential interviewer bias, I opted to exclude my own Facebook friends or anyone whom I knew personally. The pool of respondents was therefore limited to second- and third-degree connections on Facebook with which I had no previous familiarity. The recruitment message was reposted on Facebook more than 25 times, expanding the respondent pool to potentially thousands of these second- and third-degree connections. Any U.S. Facebook user over the age of 18 was eligible to participate, as long as he or she had posted a version of the red equal sign symbol as a profile picture and did not know me personally.

The resulting snowball sample of 22 respondents has obvious limitations to generalizability, although I employed the technique of maximum variation sampling to diversify the sample with regard to gender, age, race, and geographic location. Over the course of the study, 32 Facebook users volunteered to be interviewed, some of whom were purposively excluded because of demographic similarities to respondents who had already been interviewed. The resulting sample included an even distribution of 11 males and 11 females, including both a transgender male and a transgender female. There was also some success in variation by age, with the respondents ranging from 22 to 70 (with a median age of 33). In terms of geographic location, respondents resided in 12 different U.S. states as well as the District of Columbia. Creating variation in the sample with regard to race was more difficult, and ultimately only 3 of the 22 respondents identified as nonwhite (representing African American, Latino, and Asian backgrounds). In addition, almost all of the respondents reported having a high level of education, and 12 had postgraduate degrees (only 1 respondent did not currently have a college degree). The majority also claimed to be more politically active than what they perceived as average, although several claimed to be less involved than their peers. Although sexual orientation was not questioned, the respondents brought it up as part of their interview testimony (because they considered it relevant for their profile picture–changing action), and an interesting pattern emerged: almost all of the women in the sample identified as straight, and almost all of the men identified as gay. It is possible that this outcome reflects a broader pattern of participation in the red equal sign campaign on Facebook, although there are no available statistics to verify this.

The recruitment outcomes are to a large extent a reflection of the character of my own Facebook network (such as the preponderance of respondents with high levels of education, as well as members of the gay community of which I am a part), although considerable effort was made to create some variation in the sample and thus diversify the voices that were included. As in the studies outlined previously, this small-scale sample does not claim to be representative of the broader population of those who participated in the action in question and should be understood as a limited slice of a much broader population that is valuable primarily for its capacity to yield rich qualitative data and assist in the process of theory building.

The semistructured interviews were conducted over the phone over a two-month period in mid-2013, and each lasted approximately 30 to 60 minutes. Prior to being interviewed, respondents went through an informed consent process and later granted permission to use their data for future research. The confidentiality process was identical to that of the studies described previously and included the use of pseudonyms to protect the identities of respondents. The process of qualitative data analysis was also identical to that used in the earlier studies: grounded theory techniques were used to systematically code data and build conceptual categories in an inductive and iterative fashion. Portions of the data were included in an article that was published in *The Journal of Computer-Mediated Communication*.[13]

> The interview schedule consisted of the following questions (in addition to questions asking for basic demographic information):
>
> Tell me the story of changing your Facebook profile picture to a red equal sign image.
>
> What did the profile picture look like? Was it the plain red equal sign or a variation?
>
> How did you decide which red equal sign image to use, and why did you choose it?
>
> How did you initially find the red equal sign image that you chose?
>
> If you added a comment while changing your profile picture, why did you do so? What did you write, and how did you decide to write it? If you did not add a comment, why did you not?

Did your red equal sign profile picture receive any responses on Facebook? If so, what was the nature of these responses?

What do you think were your goals in changing your profile picture to a red equal sign?

Do you think these goals were met? If so, why? If not, why not?

In your opinion, what do you see as the general value of participating in the red equal sign campaign on Facebook?

What are your feelings about the Facebook red equal sign campaign more broadly? What do you think was its impact?

Are you aware of any criticism of the Facebook red equal sign campaign? If so, what do you think about this criticism?

Please tell me about your political activities more broadly. This may include participating in political organizations, attending political events and rallies, and having political conversations with others in person or online.

Generally speaking, how would you describe the strength of your political participation?

How do you see your participation in the Facebook red equal sign campaign to the political activities you just described? Do you think they are connected or separate?

Notes

Chapter 1

1. Shane DiMaio, "Home Depot Worker's 'America Was Never Great' Hat Sparks Social Media Rage," *SI Live*, May 19, 2016, accessed June 25, 2016, http://www.silive.com/news/index.ssf/2016/05/home_depot-workers_america_was_never_great_hat.html.

2. Jerome Hudson, "Home Depot Employee's 'America Was Never Great' Hat Sparks Outrage on Social Media," *Brietbart*, May 19, 2016, accessed June 25, 2016, http://www.breitbart.com/big-government/2016/05/19/home-depot-employees-america-never-great-hat-sparks-outrage-social-media/.

3. Bethania Palma Markus, "Home Depot Worker Inundated with Racist Threats for Trolling Trump with 'America Was Never Great' Hat," *Raw Story*, May 19, 2016, accessed June 25, 2016, http://www.rawstory.com/2016/05/home-depot-worker-inundated-with-racist-threats-for-trolling-trump-with-america-was-never-great-hat/.

4. Simon Sinek, "How Baby Boomers Screwed Their Kids—And Created Millennial Impatience," *Salon*, January 4, 2014, accessed October 1, 2014, http://www.salon.com/2014/01/04/how_baby_boomers_screwed_their_kids_%E2%80%94_and_created_millennial_impatience/.

5. Evgeny Morozov, "Foreign Policy: Brave New World of Slacktivism," *NPR*, May 19, 2009, accessed February 28, 2016, http://www.npr.org/templates/story/story.php?storyId=104302141.

6. Stuart Shulman, "The Case against Mass E-mails: Perverse Incentives and Low Quality Public Participation in U.S. Federal Rulemaking," *Policy & Internet* 1, no. 1 (2009): 23–53.

7. Kirk Kristofferson, Katherine White, and John Peloza, "The Nature of Slacktivism: How the Social Observability of an Initial Act of Token Support Affects Subsequent Prosocial Action," *Journal of Consumer Research* 40, no. 6 (2014): 1149–1166.

8. David Karpf, "Online Political Mobilization from the Advocacy Group's Perspective: Looking beyond Clicktivism," *Policy & Internet* 2, no. 4 (2010): 7–41.

9. Jennifer Stromer-Galley, *Presidential Campaigning in the Internet Age* (New York: Oxford University Press, 2014), 15.

10. Elihu Katz and Paul Lazarsfeld, *Personal Influence: The Part Played by People in the Flow of Mass Communications* (Glencoe, IL: Free Press, 1965).

11. W. Lance Bennett and Alexandra Segerberg, *The Logic of Connective Action: Digital Media and the Personalization of Contentious Politics* (New York: Cambridge University Press, 2013).

12. Jennifer Earl and Katrina Kimport, *Digitally Enabled Social Change: Activism in the Internet Age* (Cambridge, MA: MIT Press, 2011), 5.

13. Bennett and Segerberg, *Logic of Connective Action*, 35.

14. Ibid., 37–38.

15. Maeve Shearlaw, "Did the #BringBackOurGirls Campaign Make a Difference in Nigeria?" *Guardian*, April 14, 2015, accessed May 1, 2015, http://www.theguardian.com/world/2015/apr/14/nigeria-bringbackourgirls-campaign-one-year-on.

16. Terrence McCoy, "Michelle Obama's #BringBackOurGirls Picture Sparks Criticism of American Drone Strikes," *Washington Post*, May 15, 2014, accessed May 1, 2015, http://www.washingtonpost.com/news/morning-mix/wp/2014/05/15/michelle-obamas-bringbackourgirls-picture-sparks-criticism-of-american-drone-strikes/.

17. See Andrew Chadwick, *The Hybrid Media System: Politics and Power* (New York: Oxford University Press, 2015); Scott Wright, "Populism and Downing Street E-Petitions: Connective Action, Hybridity, and the Changing Nature of Organizing," *Political Communication* 32, no. 3 (2015): 414–433.

18. Brian Ries, "Michelle Obama: It's Time to Bring Back Our Girls," *Mashable*, May 7, 2014, accessed May 1, 2015, http://mashable.com/2014/05/07/michelle-obama-nigeria-bring-back-our-girls/#ole9EH1DaGq8.

19. For a detailed discussion of Donald Trump's use of Twitter in his 2016 presidential bid, see Michael Barbaro, "Pithy, Mean, and Powerful: How Donald Trump Mastered Twitter for 2016," *New York Times*, October 5, 2015, accessed April 5, 2016, http://www.nytimes.com/2015/10/06/us/politics/donald-trump-twitter-use-campaign-2016.html?_r=0.

20. Henry Jenkins, Sam Ford, and Joshua Green, *Spreadable Media: Creating Value and Meaning in a Networked Culture* (New York: New York University Books, 2013), 7.

21. Lee Rainie and Aaron Smith, "Politics on Social Networking Sites," *Pew Internet & American Life Project*, September 4, 2012, accessed February 1, 2015, http://pewinternet.org/Reports/2012/Politics-on-SNS.aspx.

22. Sarah Banet-Weiser, *Authentic TM: The Politics of Ambivalence in a Brand Culture* (New York: New York University Press, 2012).

23. Kalle Lasn, *Culture Jam* (New York: Eagle Brook, 1999).

24. For a critical overview of political marketing, see Nicholas O'Shaughnessy, *The Phenomenon of Political Marketing* (New York: St. Martin's Press, 1990).

25. Margaret Scammell, *Consumer Democracy: The Marketing of Politics* (Cambridge: Cambridge University Press, 2014), 177.

26. For example, see Bengu Hosch-Dayican et al., "How Do Online Citizens Persuade Fellow Voters? Using Twitter during the 2012 Dutch Parliamentary Election Campaign," *Social Science Computer Review* 34, no. 2 (2016): 135–152; Nils Gustafsson, "This Time It's Personal: Social Networks, Viral Politics and Identity Management," in *Emerging Practices in Cyberculture and Social Networking*, ed. Daniel Riha and Anna Maj (New York: Rodopi, 2010), 3–23.

27. David Karpf, *The MoveOn Effect: The Unexpected Transformation of American Political Advocacy* (New York: Oxford University Pres, 2012), 161.

28. See Jürgen Habermas, *The Theory of Communicative Action, Volume 1*, trans. Thomas McCarthy (Boston: Beacon Press, 1984).

29. Lincoln Dahlberg, "Re-Constructing Digital Democracy: An Outline of Four 'Positions,'" *New Media & Society* 13, no. 6 (2011): 860.

30. Jürgen Habermas, *Moral Consciousness and Communicative Action*, trans. Shierry Weber Nicholsen and Christian Lenhardt (Cambridge, MA: MIT Press, 1990), 89.

31. For a critical overview of the digital deliberative democracy model, see Brian Loader and Dan Mercea, "Networking Democracy? Social Media Innovations and Participatory Politics," *Information, Communication, & Society* 14, no. 6 (2011): 757–769.

32. For an overview of the "culturalist" turn, see Giovanna Mascheroni, "Performing Citizenship Online: Identity, Subactivism and Participation," *Observatorio* 7, no. 3 (2013): 93–119.

33. Peter Dahlgren, "Media, Citizenship, and Civic Culture," in *Mass Media and Society*, ed. James Curran and Michael Gurevitch (London and New York: Arnold, 2000), 310–328.

34. Maria Bakardjieva, "Subactivism: Lifeworld and Politics in the Age of the Internet," *The Information Society* 25 (2009): 103.

35. Jessica Vitak et al., "It's Complicated: Facebook Users' Political Participation in the 2008 Election," *CyberPsycholology, Behavior and Social Networking* 14, no. 3 (2011): 107–114.

36. Todd Graham, Daniel Jackson, and Scott Wright, "From Everyday Conversation to Political Action: Talking Austerity in Online 'Third Spaces,'" *European Journal of Communication* 30, no. 6 (2015): 648–665.

37. Jessica Beyer, *Expect Us: Online Communities and Political Mobilization* (New York: Oxford University Press, 2014).

38. Shelley Boulianne, "Social Media Use and Participation: A Meta-Analysis of Current Research," *Information, Communication, and Society* 18, no. 5 (2015): 524–538.

39. Josh Pasek, Eian More, and Daniel Romer, "Realizing the Social Internet? Online Social Networking Meets Offline Civic Engagement," *Journal of Information Technology & Politics* 6, no. 3–4, (2009): 197–215.

40. Boulianne, "Social Media Use," 529.

41. Joakim Ekman and Erik Amna, "Political Participation and Civic Engagement: Towards a New Typology," *Human Affairs* 22, no. 3 (2012): 289.

42. Ibid., 292.

43. Ibid.

44. Manuel Castells, "Communication, Power, and Counter-power in the Network Society," *International Journal of Communication* 1 (2007): 249.

45. Ibid., 250.

46. Kevin DeLuca and Jennifer Peeples, "From Public Sphere to Public Screen: Democracy, Activism, and the 'Violence' of Seattle," *Critical Studies in Media Communication* 19, no. 2 (2002): 144.

47. Chantal Mouffe, "Deliberative Democracy or Agonistic Pluralism," *Reihe Politikwissenschaft/ Political Science Series* 72 (2000): 17, accessed February 1, 2016, https://www.ihs.ac.at/publications/pol/pw_72.pdf.

48. However, the persuasion framework considered here does not necessarily adhere to the normative dimension of Mouffe's theory, which advocates pluralism and tolerance and argues that adversaries in democratic conflict must still share common values of liberty and equality.

49. Zizi Papacharissi, *A Private Sphere: Democracy in a Digital Age* (Malden, MA: Polity Press, 2010), 119.

50. Daniel Kreiss, *Taking Our Country Back: The Crafting of Networked Politics from Howard Dean to Barack Obama* (New York: Oxford University Press, 2012), 26.

51. Steven Best and Douglas Kellner, *The Postmodern Turn* (New York: Guilford Press, 1997), 177.

52. Kevin Barnhurst, "Politics in the Fine Meshes: Young Citizens, Power, and Media," *Media, Culture, and Society* 20, no. 2 (1998): 209.

53. John Gibbins and Bo Reimer, *The Politics of Postmodernity: An Introduction to Contemporary Politics and Culture* (London: Sage, 1999), 113.

54. Lasn, *Culture Jam*, 123.

55. Ibid., 132.

56. Douglas Rushkoff, *Media Virus! Hidden Agendas in Popular Culture* (New York: Ballantine Books, 1994), 8.

57. Lasn, *Culture Jam*, 212.

58. For a discussion of the influence of Rushkoff's book on the viral media model, see Henry Jenkins, "If It Doesn't Spread, It's Dead (Part One): Media Viruses and Memes," *Henryjenkins.org*, February 11, 2009, accessed March 15, 2015, http://henryjenkins.org/2009/02/if_it_doesnt_spread_its_dead_p.html.

59. Rushkoff, *Media Virus!*, 15.

60. Jenkins, "If It Doesn't Spread."

61. Jenkins, Ford, and Green, *Spreadable Media*.

62. Bennett and Segerberg, *Logic of Connective Action*, 37–38.

63. Zizi Papacharissi, *Affective Publics: Sentiment, Technology, and Politics* (New York: Oxford University Press, 2014), 35.

64. Mary Douglas, *Purity and Danger* (New York: Praeger, 1966).

65. Papacharissi, *Affective Publics*, 28.

66. Ibid., 130.

67. Andrew Wernick, *Promotional Culture: Advertising, Ideology, and Symbolic Expression* (London: Sage, 1991), 182.

68. Quoted in Vance Packard, *The Hidden Persuaders* (New York: David McKay, 1957), 200.

69. Joe McGinnis, *The Selling of the President 1968* (New York: Trident Press, 1969).

70. Kathleen Hall Jamieson, *Packaging the Presidency: A History and Criticism of Presidential Campaign Advertising* (New York: Oxford University Press, 1984).

71. Elaine Thompson, "Political Culture," in *Americanisation and Australia*, ed. Paul Bell and Roger Bell (Sydney: University of New South Wales Press, 1998), 120.

72. John Street, *Politics and Popular Culture* (Philadelphia: Temple University Press, 1997), 45–62.

73. For example, see Sut Jhally, *The Codes of Advertising: Fetishism and the Political Economy of Meaning in the Consumer Society* (New York: Routledge, 1990).

74. Scammell, *Consumer Democracy*, 21.

75. Ibid., 15–16.

76. Nicholas O'Shaughnessy, "The Death and Life of Propaganda," *Journal of Public Affairs* 12, no. 1 (2012): 29–36.

77. Scammell, *Consumer Democracy*, 14.

78. Stuart Ewen, *All Consuming Images: The Politics of Style in Contemporary Culture, Revised Edition* (New York: Basic Books, 1999), 269–271.

79. See Barrie Axford and Richard Huggins, "Anti-politics or the Triumph of Postmodern Populism in Promotional Cultures?" *Javnost: The Public* 4, no. 3 (1997): 5–27; Richard Huggins, "The Transformation of the Political Audience?" In *New Media and Politics*, ed. Barrie Axford and Richard Huggins (London: Sage, 2001), 127–150.

80. Barrie Axford, "The Transformation of Politics or Anti-Politics?," in *New Media and Politics*, ed. Barrie Axford and Richard Huggins (London: Sage, 2001), 7.

81. Scammell, *Consumer Democracy*, 34.

82. Justin Lewis, Sanna Inthorn, and Karin Wahl-Jorgensen, *Citizens or Consumers?: What the Media Tell Us about Political Participation* (New York: Open University Press, 2005), 138.

83. Nico Carpentier, *Media and Participation: A Site of Ideological-Democratic Struggle* (Bristol, UK: Intellect, 2011), 84.

84. Liz Cohen, *A Consumer's Republic: The Politics of Mass Consumption* (New York: Knopf, 2003), 347.

85. Heather Savigny, *The Problem of Political Marketing* (New York: Continuum, 2008), 116.

86. Erik Asard and W. Lance Bennett, *Democracy and the Marketplace of Ideas: Communication and Government in Sweden and the United States* (Cambridge: Cambridge University Press, 1997), 47.

87. Alison Hearn, "Brand Me 'Activist,'" in *Commodity Activism: Cultural Resistance in Neoliberal Times*, ed. Roopali Mukherjee and Sarah Banet-Weiser (New York: New York University Press, 2012), 23–38.

88. Peter Dahlgren, *The Political Web: Media, Participation, and Alternative Democracy* (New York: Palgrave, 2013), 52.

89. Ibid., 63.

90. Ibid., 52.

91. Cohen, *Consumer's Republic*, 8.

92. W. Lance Bennett, "Changing Citizenship in the Digital Age," in *Civic Life Online: Learning How Digital Media Can Engage Youth*, ed. Lance W. Bennett (Cambridge, MA: MIT Press, 2008), 1–24.

93. Ibid., 14.

94. Nestor Garcia Canclini, *Consumers and Citizens: Globalization and Multicultural Conflicts* (Minneapolis: University of Minnesota Press, 2001), 5.

95. W. Lance Bennett, "Branded Political Communication: Lifestyle Politics, Logo Campaigns, and the Rise of Global Citizenship," in *The Politics behind Products*, ed. Michele Micheletti, Andreas Follesdal, and Dietlind Stolle (New Brunswick, NJ: Transaction Books, 2008), 101–126.

96. Jamie Warner, "Political Culture Jamming: The Dissident Humor of 'The Daily Show with Jon Stewart,'" *Popular Communication* 5, no. 1 (2007): 18–19.

97. Mark Dery, "The Merry Pranksters and the Art of the Hoax," *New York Times*, December 23, 1990, accessed September 12, 2015, http://www.nytimes.com/1990/12/23/arts/the-merry-pranksters-and-the-art-of-the-hoax.html?pagewanted=all.

98. Warner, "Political Culture Jamming," 19.

99. W. Glynn Mangold and David Faulds, "Social Media: The New Hybrid Element of the Promotion Mix," *Business Horizons* 52, no. 4 (2009): 357–365.

100. Banet-Weiser, *Authentic TM*, 7.

101. Jenkins, Ford, and Green, *Spreadable Media*, 7.

102. See Mangold and Faulds, "The New Hybrid Element."

103. Rob Fuggetta, *Brand Advocates: Turning Enthusiastic Customers into a Powerful Marketing Force* (Hoboken, NJ: Wiley, 2012).

104. Alex Goldfayn, *Evangelist Marketing: What Apple, Amazon, and Netflix Understand about Their Customers (That Your Company Probably Doesn't)* (Dallas: BenBella Books, 2013).

105. Jackie Huba and Ben McConnell, *Citizen Marketers: When People Are the Message* (Chicago: Kaplan Business, 2007), 4.

106. Ruth E. Brown, "Citizen Marketing," in *Handbook of Research on Social Interaction Technologies and Collaboration Software: Concepts and Trends: Concepts and Trends*, ed. Tatyana Dumova (Hershey, PA: IGI Global, 2010), 45.

107. Jenkins, Ford, and Green, *Spreadable Media*, 7.

108. Michael Serazio, "Managing the Digital News Cyclone: Power, Participation, and Political Production Strategies," *International Journal of Communication* 9 (2015): 1919.

109. Stuart Hall, "What Is This 'Black' in Black Popular Culture?," in *Stuart Hall: Critical Dialogues in Cultural Studies*, ed. David Morley and Kuan-Hsing Chen (London: Routledge, 1996), 470.

110. Limor Shifman, "Memes in a Digital World: Reconciling with a Conceptual Troublemaker," *Journal of Computer-Mediated Communication* 18, no. 3 (2013): 365.

111. See Steven Rosenbaum, *Curation Nation: How to Win in a World Where Consumers Are Creators* (New York: McGraw–Hill Professional, 2011).

112. Jenkins, "If It Doesn't Spread."

113. Eytan Bakshy, "Showing Support for Marriage Equality on Facebook," *Facebook Data Science*, March 27, 2013, accessed October 1, 2014, https://www.facebook.com/notes/facebook-data-science/showing-support-for-marriage-equality-on-facebook/10151430548593859.

114. For example, see Jennifer Lees-Marshment, "The Marriage of Politics and Marketing," *Political Studies* 49, no. 4 (2001): 692–713.

Chapter 2

1. Roger A. Fischer, *Tippecanoe and Trinkets Too: The Material Culture of American Presidential Campaigns, 1828–1984* (Urbana and Chicago: University of Illinois Press, 2008), 33.

2. Ibid., 235.

3. Michael Schudson, *The Good Citizen: A History of American Civic Life* (Cambridge, MA: Harvard University Press, 1998).

4. Kathleen Hall Jamieson, *Packaging the Presidency: A History and Criticism of Presidential Campaign Advertising* (New York: Oxford University Press, 1984), 17.

5. Henry Jenkins, *Convergence Culture: When Old and New Media Collide* (New York: New York University Press, 2006), 135–137.

6. For an extended discussion of the "culturalization" of politics, see John Corner and Dick Pels, "Introduction: The Re-Styling of Politics," in *Media and the Restyling of Politics: Consumerism, Celebrity, and Cynicism*, ed. John Corner and Dick Pels (London: Sage, 2003), 1–18.

7. danah boyd, "Why Youth (Heart) Social Network Sites: The Role of Networked Publics in Teenage Social Life," in *Youth, Identity, and Digital Media*, ed. David Buckingham (Cambridge, MA: MIT Press, 2008), 129.

8. See Philip M. Taylor, *Munitions of the Mind: A History of Propaganda from the Ancient World to the Present Day*, 3rd ed. (Manchester, UK: Manchester University Press, 2003).

9. Emile Durkheim, *The Elementary Forms of Religious Life*, trans. J. W. Swain (New York: Free Press, 1965), 230.

10. Marcel Danesi, *Why It Sells: Decoding the Meanings of Brand Names, Logos, Ads, and Other Marketing and Advertising Ploys* (New York: Rowman & Littlefield, 2007), 93.

11. Howard S. Levy, "Yellow Turban Religion and Rebellion at the End of Han," *Journal of the American Oriental Society* 76, no. 4 (1956): 214–227.

12. See Thomas Woodcock and John Martin Robinson, *The Oxford Guide to Heraldry* (Oxford: Oxford University Press, 1988).

13. John R. Milton, *Areopagitica* (Oxford: Clarendon Press, 1886), 51–52.

14. Margaret Scammell, *Consumer Democracy: The Marketing of Politics* (Cambridge: Cambridge University Press, 2014), 180.

15. For instance, Jürgen Habermas makes such an argument in his widely cited theory of the public sphere, which celebrates the role of 18th-century European bourgeois print culture (along with bourgeois salons) in supporting an active and informed democratic citizenry. Jürgen Habermas, *Structural Transformation of the Public Sphere*, trans. Thomas Burger (Cambridge, MA: MIT Press, 1989).

16. See Bob Harris, *Politics and the Rise of the Press: Britain and France, 1620–1800* (London: Routledge, 1996).

17. Geoffrey Lock, "The 1689 Bill of Rights," *Political Studies* 37, no. 14 (1989): 540–561.

18. Jennny Uglow, *William Hogarth: A Life and a World* (London: Faber & Faber, 2010), 965–977.

19. Fischer, *Tippecanoe and Trinkets Too*, 15.

20. Nicholas O'Shaughnessy, *The Phenomenon of Political Marketing* (New York: St. Martin's Press, 1990), 39.

21. Fischer, *Tippecanoe and Trinkets Too*, 18.

22. Charles Tilly, *Popular Contention in Great Britain, 1758–1834* (Cambridge, MA: Harvard University Press, 1995), 45–47.

23. Ibid., 61.

24. Ibid., 57.

25. Ibid., 374.

26. Ibid.

27. Susan Davis, *Parades and Power: Street Theatre in Nineteenth-Century Philadelphia* (Berkeley: University of California Press, 1988), 159.

28. Ibid., 158.

29. Ibid.

30. Jamieson, *Packaging the Presidency*, 17.

31. John Thompson, *The Media and Modernity: A Social Theory of the Media* (Cambridge, MA: Polity, 1995), 126.

32. See Kitty Kelley and Stanley Tretick, *Let Freedom Ring: Stanley Tretick's Iconic Images of the March on Washington* (New York: Macmillan, 2013).

33. Fischer, *Tippecanoe and Trinkets Too*, 234.

34. Ibid., 236.

35. Shay Sayre, "T-shirt Messages: Fortune or Folly for Advertisers?," in *Advertising and Popular Culture: Studies in Variety and Versatility*, ed. Sammy Richard Danna (Bowling Green, OH: Bowling Green State University Popular Press, 1992), 73.

36. Wendy Parkins, "'The Epidemic of Purple, White, and Green': Fashion and the Suffragette Movement in Britain 1908–1914," in *Fashioning the Body Politic: Gender, Dress, Citizenship*, ed. Wendy Parkins (Oxford: Berg, 2002), 99.

37. Quoted in Marian Sawer, "Wearing Your Politics on Your Sleeve: The Role of Political Colours in Social Movements," *Social Movement Studies: Journal of Social, Cultural, and Political Protest* 6, no. 1 (2007): 47.

38. Lisa Tickner, *The Spectacle of Women: Imagery of the Suffrage Campaign 1907–14* (Chicago: University of Chicago Press, 1988), 294.

39. Diane Crane, *Fashion and Its Social Agendas: Class, Gender, and Identity in Clothing* (Chicago: University of Chicago Press, 2000).

40. Anthony Giddens, *Modernity and Self-Identity: Self and Society in the Late Modern Age* (Stanford, CA: Stanford University Press, 1991).

41. Elizabeth Wilson, *Adorned in Dreams: Fashion and Modernity* (New Brunswick, NJ: Rutgers University Press), 184.

42. Ibid., 198.

43. Dick Hebdige, *Subculture: The Meaning of Style* (London: Methuen, 1979).

44. Ernest Hamburger, "Contrasting the Hippie and Junkie," *The International Journal of the Addictions* 4, no. 1 (1969): 129.

45. Quoted in Ian Inglis, "The Continuing Story of John Lennon," *Critical Studies in Media Communication* 22, no. 5 (2005): 452.

46. This photograph was later used by the clothing company Worn Free to create a mass-market reproduction T-shirt with the "Come Together" slogan and peace sign symbol.

47. J. D. Reed, "Hail to the T, the Shirt That Speaks Volumes," *Smithsonian* 23, no. 1 (1992): 96–102.

48. See Charlotte Brunel, *The T-shirt Book* (New York: Assouline, 2002), 111.

49. See Steve Crist and Howard L. Bingham, *Howard L. Bingham's Black Panthers 1968* (New York: Ammo Books, 2009).

50. Patricia Brennan, "Making a Statement," *Washington Post*, October 10, 1982, 47.

51. Crane, *Fashion and Its Social Agendas*, 177.

52. Neil Fisher, "'Free Angela Davis' Shirt," *The Times* (London), October 6, 2004, 16.

53. Sheryl Garratt, "We Love T-shirts," *The Times* (London), June 22, 2002, Features.

54. Larry Gross, *Up from Invisibility: Lesbians, Gay Men, and the Media in America* (New York: Columbia University Press, 2001), 19.

55. See Dennis Williams, "Homosexuals: Anita Bryant's Crusade," *Newsweek*, April 11, 1977, 39.

56. Paul Taylor, *Impresario: Malcolm McLaren and the British New Wave* (New York: New Museum of Contemporary Art, 1988).

57. Bernadine Morris, "British Fashion Rises and Shines," *New York Times*, March 20, 1984, B14.

58. Kate Saunders, "Fashionable Causes," *Sunday Times* (London), November 18, 1990, Features.

59. Susannah Frankel, "Been There, Done That, Made the T-shirt," *Independent* (London), September 19, 2001, 8.

60. Nina Hyde, "T-time at 10 Downing St: Margaret Thatcher's Big Boost for Fashion," *Washington Post*, March 19, 1984, C1.

61. Harriet Lane, "Been There. Sold Out. Want the T-shirt?: Katharine Hamnett, One Time Left-wing Sloganeer, Is on a New Crusade," *Observer (London)*, February 14, 1999, 4.

62. Quoted in Irene Lacher, "Suited to a T," *Los Angeles Times*, October 10, 1993, accessed September 5, 2014, http://articles.latimes.com/1993-10-10/news/vw-44413_1_t-shirt-sales.

63. Saunders, "Fashionable Causes."

64. John D. Battersby, "South Africa Bars Protests Backing Detainee Freedom," *New York Times*, April 12, 1987, A1.

65. Nicholas D. Kristof, "Even Gloomy T-Shirts Fall under Censorship," *New York Times*, July 29, 1991, A4.

66. Sally Young, *The Persuaders: Inside the Hidden Machine of Political Advertising* (Melbourne: Pluto Press Australia, 2004), 17.

67. Lyman Chaffee, *Political Protest and Street Art: Popular Tools for Democratization in Hispanic Countries* (Westport, CT: Greenwood Press, 1993), 141.

68. Richard Martin and Harold Koda, "V-O-T-E: The 1992 Election in T-shirts," *Textile and Text* 14, no. 3 (1992): 30–35.

69. Ibid., 30.

70. James Endersby and Michael Towle, "Tailgate Partisanship: Political and Social Expression through Bumper Stickers," *Social Science Journal* 33, no. 3 (1996): 319.

71. Douglas Rushkoff, *Media Virus! Hidden Agendas in Popular Culture* (New York: Ballantine Books, 1994), 241.

72. Ibid., 242.

73. Ibid., 243.

74. Ibid., 242–243.

75. For a discussion of Richard Dawkins's work on memes and its relation to Internet meme culture, see Limor Shifman, "Memes in a Digital World: Reconciling with a Conceptual Troublemaker," *Journal of Computer-Mediated Communication* 18, no. 3 (2013): 362–377.

76. Rushkoff, *Media Virus!*, 249–253.

77. W. Lance Bennett, "New Media Power," in *Contesting Media Power*, ed. Nick Couldry and John Curran (Lanham, MD: Rowman & Littlefield, 2003), 32–33.

78. Ibid.

79. Howard Rheingold, *Smart Mobs: The Next Social Revolution* (Cambridge, MA: Basic Books, 2002), 158–160.

80. Jennifer Earl and Katrina Kimport, *Digitally Enabled Social Change: Activism in the Internet Age* (Cambridge, MA: MIT Press, 2011).

81. Rheingold, *Smart Mobs*, 158.

82. Mike Giglio, quoted in Sahar Khamis and Katherine Vaughn, "Cyberactivism in the Egyptian Revolution: How Civic Engagement and Citizen Journalism Tilted the Balance," *Arab Media and Society* 13, no. 3 (2011): 10.

83. Earl and Kimport, *Digitally Enabled Social Change*, 9.

84. "We the People," accessed August 2, 2014, https://petitions.whitehouse.gov/.

85. David Karpf, "Online Political Mobilization from the Advocacy Group's Perspective: Looking beyond Clicktivism," *Policy & Internet* 2, no. 4 (2010): 7–41.

86. Stuart Shulman, "The Case against Mass E-mails: Perverse Incentives and Low Quality Public Participation in U.S. Federal Rulemaking," *Policy & Internet* 1, no. 1 (2009): 23–53.

87. Scott Wright, "'Success' and Online Political Participation: The Case of Downing Street E-petitions," *Information, Communication, & Society* 19, no. 6 (2016): 843–857.

88. Karpf, "Online Political Mobilization," 9.

89. Henrik Christensen, "Political Activities on the Internet: Slacktivism or Political Participation by Other Means?" *First Monday* 16, no. 2 (2011), accessed February 15, 2016, http://dx.doi.org/10.5210/fm.v16i2.3336.

90. Monte Solberg, "Real #Change Doesn't Happen on Twitter," *Toronto Sun*, June 1, 2014, accessed October 10, 2014, http://www.torontosun.com/2014/05/30/real-change-doesnt-happen-on-twitter.

91. Asher Wren, "Why #BringBackOurGirls Isn't Just Another Slacktivism Campaign," *Huffington Post UK*, May 28, 2014, accessed October 10, 2014, http://www.huffington-post.co.uk/asher-wren/bringbackourgirls-twitter-campaign_b_5399725.html.

Chapter 3

1. For a summary of the red equal sign meme's many variations, see "HRC Goes Viral," Human Rights Campaign, accessed August 1, 2014, http://www.hrc.org/viral.

2. Although the Facebook Data Science Team was unable to verify the exact number of users who posted a red equal sign image, the group came to the 2.7 million estimate by measuring the increase in profile picture changes on Tuesday, March 26, 2013, compared to the previous Tuesday. See Eytan Bakshy, "Showing Support for Marriage Equality on Facebook," *Facebook Data Science*, March 29, 2013, accessed October 1, 2014, https://www.facebook.com/notes/facebook-data-science/showing-support-for-marriage-equality-on-facebook/10151430548593859.

3. Anastasia Khoo, "Painting the Internet Red," *Stanford Social Innovation Review*, May 1, 2013, accessed October 1, 2014, http://www.ssireview.org/blog/entry/painting_the_internet_red.

4. Melanie Tannenbaum, "Will Changing Your Profile Picture Do Anything for Marriage Equality?" *Scientific American*, March 28, 2013, accessed April 4, 2015, http://blogs.scientificamerican.com/psysociety/2013/03/28/marriage-equality-and-social-proof/.

5. Brian Moylan, "The Red Marriage Equality Sign on Your Facebook Profile Is Completely Useless," *Vice*, March 26, 2013, accessed April 4, 2015, http://www.vice.com/read/the-red-marriage-equality-sign-on-your-facebook-profile-is-completely-useless.

6. Alex Fitzpatrick, "Did the Red Equal Sign Avatars Affect Court's Marriage Decisions?" *Mashable*, June 26, 2013, accessed April 4, 2015, http://mashable.com/2013/06/26/red-equal-sign-avatars/.

7. Naomi Klein, *No Logo* (New York: Picador, 2000), 44.

8. *Crumb*, directed by Terry Zwigoff (1994; New York: Criterion Collection, 2010), DVD.

9. Klein, *No Logo*, 28–30.

10. See Mark Peters, "Virtue Signaling and Other Inane Platitudes," *The Boston Globe*, December 24, 2015, accessed May 10, 2016, https://www.bostonglobe.com/ideas/2015/12/24/virtue-signaling-and-other-inane-platitudes/YrJRcvxYMofMcCfgORUcFO/story.html.

11. Yifat Mor, Neta Kligler-Vilenchik, and Ifat Maoz, "Political Expression on Facebook in a Context of Conflict: Dilemmas and Coping Strategies of Jewish–Israeli Youth," *Social Media + Society* 1, no. 2 (2015), accessed March 15, 2015, doi:10.1177/2056305115606750.

12. Peter Dahlgren, *The Political Web: Media, Participation, and Alternative Democracy* (New York: Palgrave Macmillan, 2013), 63.

13. Zizi Papacharissi, *A Private Sphere: Democracy in a Digital Age* (Malden, MA: Polity Press, 2010), 143.

14. Ibid., 144.

15. Linda Martín Alcoff, *Visible Identities: Race, Gender, and the Self* (New York: Oxford University Press, 2005).

16. Hannah Arendt, *The Human Condition* (Chicago: University of Chicago Press, 1958), 357.

17. Ibid., 195.

18. Michael Warner, "Publics and Counterpublics," *Public Culture* 14, no. 1 (2002): 61–63.

19. For an account of Habermas's ideal of the rational public sphere of deliberation and consensus building, see Jürgen Habermas, *The Theory of Communicative Action, Volume 1*, trans. Thomas McCarthy (Boston: Beacon Press, 1984).

20. Andrea Brighenti, *Visibility in Social Theory and Social Research* (New York: Palgrave Macmillan, 2010).

21. Lyman Chaffee, *Political Protest and Street Art: Popular Tools for Democratization in Hispanic Countries* (Westport, CT: Greenwood Press, 1993), 3.

22. Ibid., 15.

23. Ernesto Laclau, *Politics and Ideology in Marxist Theory* (London: New Left Books, 1977), 167–173.

24. Lauren Berlant and Elizabeth Freeman, "Queer Nationality," in *Fear of a Queer Planet*, ed. Michael Warner (Minneapolis: University of Minnesota Press, 1993), 201.

25. For an in-depth discussion of gay and lesbian visibility in the mass media, see Larry Gross, *Up from Invisibility* (New York: Columbia University Press, 2001).

26. Charles Morris and John Sloop, "'What Lips These Lips Have Kissed': Refiguring the Politics of Queer Public Kissing," *Communication and Critical/Cultural Studies* 3, no. 1 (2006): 23.

27. Berlant and Freeman, "Queer Nationality," 201–207.

28. Kevin DeLuca, "Unruly Arguments: The Body Rhetoric of Earth First!, ACT UP, and Queer Nation," *Argumentation and Advocacy* 36, no. 1 (1999): 18.

29. W. Lance Bennett and Alexandra Segerberg, *The Logic of Connective Action: Digital Media and the Personalization of Contentious Politics* (New York: Cambridge University Press, 2013).

30. Suzanna Walters, *All the Rage: The Story of Gay Visibility in America* (Chicago: University of Chicago Press, 2001).

31. Tina Chanter, "A Critique of Martín Alcoff's Identity Politics: On Power and Universality," *Philosophy Today* 53 (2009): 50.

32. Morris and Sloop, "What Lips These Lips Have Kissed," 23.

33. For a detailed description of this concept, see Robert Cialdini, "Harnessing the Science of Persuasion," *Harvard Business Review* 79 (2001): 72–79.

34. For a journalistic account of campus-based T-shirt visibility campaigns, see Kate Zernike, "Fighting Homophobia? A Fine Idea," *New York Times*, January 18, 2004, A7.

35. Elihu Katz and Paul Lazarsfeld, *Personal Influence: The Part Played by People in the Flow of Mass Communications* (Glencoe, IL: Free Press, 1965).

36. Henry Jenkins, "If It Doesn't Spread, It's Dead (Part One): Media Viruses and Memes," *Henryjenkins.org*, February 11, 2009, accessed March 15, 2015, http://henryjenkins.org/2009/02/if_it_doesnt_spread_its_dead_p.html.

37. Anastasia Khoo, "How Support for Marriage Equality Went Viral on Facebook," *The Huffington Post*, June 22, 2013, accessed October 1, 2014, http://www.huffington-post.com/anastasia-khoo/how-support-for-marriage-_b_3483736.html?utm_hp_ref=impact&ir=Impact.

38. "What Is the It Gets Better Project?," Itgetsbetter.org, accessed October 13, 2014, http://www.itgetsbetter.org/pages/about-it-gets-better-project/.

39. Brian Stelter, "Campaign Offers Help to Gay Youths," *The New York Times*, October 18, 2010, accessed October 4, 2014, http://www.nytimes.com/2010/10/19/us/19video.html?_r=0.

40. Dahlgren, *The Civic Web*, 55.

41. Bill Wasik, "#Riot: Self-Organized, Hyper-Networked Revolts, Coming to a City Near You," *Wired*, December 16, 2011, accessed October 20, 2014, http://www.wired.com/magazine/2011/12/ff_riots/.

42. Henry Jenkins, Mizuko Ito, and danah boyd, *Participatory Culture in a Networked Era: A Conversation on Youth, Learning, Commerce, and Politics* (New York: Polity Press, 2015), 175.

43. Eric Adler, Mara Rose Williams, and Kaitlyn Klein, "After California Killings, Twitter's #YesAllWomen Reveals the Vast Extent of Misogyny," *The Kansas City Star*, June 3, 2014, accessed March 20, 2015, http://www.kansascity.com/news/local/article438881/

After-California-killings-Twitter%E2%80%99s-YesAllWomen-reveals-the-vast-extent-of-misogyny.html.

44. Emmanuella Grinberg, "Meredith Vieira Explains #WhyIStayed," *CNN*, September 17, 2014, accessed March 20, 2015, http://www.cnn.com/2014/09/09/living/rice-video-why-i-stayed/.

45. Logan Rhodes and Adrian Carrasquillo, "How the Powerful #IfTheyGunnedMeDown Movement Changed the Conversation about Michael Brown's Death," *Buzzfeed*, August 13, 2014, accessed March 20, 2015, http://www.buzzfeed.com/mrloganrhoades/how-the-powerful-iftheygunnedmedown-movement-changed-the-con.

46. Wasik, "#Riot."

47. "We Are the 99 Percent," accessed October 13, 20104, http://wearethe99percent.tumblr.com/.

48. Jodi Dean, "Claiming a Division, Naming a Wrong," *Theory & Event* 14, no. 4 (2011): supplement.

49. Ethan Zuckerman, "Understanding Digital Civics," *My Heart's in Accra*, August 30, 2012, accessed May 1, 2015, http://www.ethanzuckerman.com/blog/2012/08/30/understanding-digital-civics/.

50. Walters, *All the Rage*, 24.

51. Amy Gluckman and Betsy Reed, *Homo Economics: Capitalism, Community, and Lesbian and Gay Life* (New York: Routledge, 1997), 17.

52. Katherine Sender, *Business, Not Politics: The Making of the Gay Market* (New York: Columbia University Press, 2004), 166.

53. See Alexandra Chasin, *Selling Out: The Gay and Lesbian Movement Goes to Market* (New York: Palgrave, 2000).

54. Sarah Warner, *Acts of Gaiety: LGBT Performance and the Politics of Pleasure* (Ann Arbor: University of Michigan Press, 2012), xi.

55. Derrick Clifton, "What's Behind Criticisms of Those Red Equal Signs in Your Facebook Feed?," *The Huffington Post*, March 29, 2013, accessed October 13, 2014, http://www.huffingtonpost.com/derrick-clifton/human-rights-campaign-same-sex-marriage_b_2973131.html.

56. Jenkins, "If It Doesn't Spread."

57. Richard L. Edwards and Chuck Tryon, "Political Video Mashups as Allegories of Citizen Empowerment," *First Monday* 14, no. 10 (2009), accessed February 2, 2016, http://dx.doi.org/10.5210/fm.v14i10.2617.

58. For a discussion of digital malleability in the context of internet memes, see Patrick Davison, "The Language of Internet Memes," in *The Social Media Reader*, ed. Michael Mandiberg (New York: New York University Press, 2012), 120–135.

59. Lance W. Bennett, "New Media Power: The Internet and Global Activism," in *Contesting Media Power: Alternative Media in a Networked World*, ed. Nick Couldry and James Curran (London: Rowman & Littlefield, 2003), 34.

60. Anastasia Khoo, "Painting the Internet Red."

61. W. Lance Bennett and Alexandra Segerberg, *The Logic of Connective Action: Digital Media and the Personalization of Contentious Politics* (New York: Cambridge University Press, 2013).

62. Natalie Fenton and Veronica Barassi, "Alternative Media and Social Networking Sites: The Politics of Individuation and Political Participation," *The Communication Review* 14, no. 3 (2011): 181–183.

63. In fact, in a Google search, I was only able to locate a single mainstream news article that mentioned the greater than profile picture image as a critical variation of the campaign. The authors briefly note toward the end of the piece in question that those who had posted it believe that "inequality will continue . . . even if the high court strikes down California's ban on gay marriage." See Matt Stevens and Jessica Garrison, "Pro-Gay Marriage Red Logo Goes Viral on Social Media," *Los Angeles Times*, March 26,

2013, accessed September 4, 2015, http://articles.latimes.com/2013/mar/26/local/la-me-ln-gay-marriage-red-logo20130326.

64. Limor Shifman, *Memes in Digital Culture* (Cambridge, MA: MIT Press, 2013), 12.

65. Davison, "The Language of Internet Memes," 123.

66. In the context of digital composition, this notion of content being deliberately designed to move quickly from peer to peer and be shared by others has been referred to as "rhetorical velocity." See David Sheridan, Jim Ridolfo, and Anthony Michel, *The Available Means of Persuasion* (Anderson, SC: Parlor, 2012), 79.

67. Kimberlé Williams Crenshaw, "Mapping the Margins: Intersectionality, Identity Politics, and Violence against Women of Color," *Stanford Law Review* 43, no. 6 (1991): 1241–1299.

Chapter 4

1. Hannah Jewell, "Ed Miliband Has Developed a Small but Growing Fandom of Teen Girls," *Buzzfeed*, April 21, 2015, accessed February 11, 2016, http://www.buzzfeed.com/hannahjewell/the-milifandom#.pcg1gQ62ko.

2. Jake Tapper, "Music Video Has a 'Crush on Obama,'" *ABC News*, June 13, 2007, accessed February 11, 2016, http://abcnews.go.com/Politics/story?id=3275802&page=1.

3. Michael Wilkinson, "Never Mind #Milifandom, Meet the #Cameronettes Who Can't Get Enough of David Cameron," *Telegraph*, April 22, 2015, accessed February 11, 2016, http://www.telegraph.co.uk/news/politics/david-cameron/11554249/Never-mind-milifandom-meet-the-cameronettes-who-cant-get-enough-of-David-Cameron.html.

4. Stuart Heritage, "#JezWeCan: Why Jeremy Corbyn Gets the Social Media Vote," *Guardian*, August 4, 2015, accessed February 11, 2016, http://www.theguardian.com/politics/2015/aug/04/jezwecan-jeremy-corbyn-social-media-vote-labour-leadership.

5. Gregory Krieg, "How This Hashtag Is Helping to Grow Bernie Sanders' Powerful Grassroots Movement," *Mic*, July 30, 2015, accessed February 12, 2016, http://mic.com/articles/123168/how-this-hashtag-is-helping-to-grow-bernie-sanders-powerful-grass-roots-movement#.uTARZ0fsa.

6. Daniella Diaz, "#BabiesForBernie Catches on Social Media," *CNN*, October 17, 2015, accessed February 12, 2016, http://www.cnn.com/2015/10/17/politics/babies-for-bernie-sanders-facebook-instagram/index.html.

7. Richard Iton, *In Search of the Black Fantastic: Politics and Popular Culture in the Post-Civil Rights Era* (Oxford: Oxford University Press, 2008).

8. Kathleen Hall Jamieson, *Packaging the Presidency: A History and Criticism of Presidential Campaign Advertising* (New York: Oxford University Press, 1984), 17.

9. Andrew Chadwick, *The Hybrid Media System: Politics and Power* (New York: Oxford University Press, 2015).

10. For instance, the phrase "your consumer is your marketer" is offered by a digital marketing professional in the PBS *Frontline* documentary "Generation Like," produced by Frank Koughan and Douglas Rushkoff (2014), accessed March 1, 2016, http://www.pbs.org/wgbh/frontline/film/generation-like/.

11. Daniel Kreiss, *Taking Our Country Back: The Crafting of Networked Politics from Howard Dean to Barack Obama* (New York: Oxford University Press, 2012), 8.

12. Jennifer Stromer-Galley, *Presidential Campaigning in the Internet Age* (New York: Oxford University Press, 2014), 169.

13. Ibid., 5.

14. Michael Serazio, "Managing the Digital News Cyclone: Power, Participation, and Political Production Strategies," *International Journal of Communication* 9 (2015): 1921.

15. Rachel K. Gibson, "Party Change, Social Media and the Rise of 'Citizen-Initiated Campaigning." *Party Politics* 21, no. 2 (2015): 190.

16. Christopher Hunter, "Political Privacy and Online Politics: How E-Campaigning Threatens Voter Privacy," *First Monday* 7, no. 2 (2002), accessed August 5, 2015, http://firstmonday.org/issues/issue7_2/hunter/index.html.

17. Daniel Kreiss and Philip N. Howard, "New Challenges to Political Privacy: Lessons from the First US Presidential Race in the Web 2.0 Era," *International Journal of Communication* 4 (2010): 1033.

18. Stromer-Galley, *Presidential Campaigning*, 18.

19. Kreiss, *Taking Our Country Back*, 194.

20. Ibid., 196–197.

21. Gibson, "Party Change," 183.

22. Jessica Baldwin-Philippi, *Using Technology, Building Democracy: Digital Campaigning and the Construction of Citizenship* (New York: Oxford University Press, 2015), 5.

23. See Margaret Scammell, *Consumer Democracy: The Marketing of Politics* (Cambridge: Cambridge University Press, 2014), 24–30.

24. John Street, "The Celebrity Politician: Political Style and Popular Culture," In *Media and the Restyling of Politics: Consumerism, Celebrity, and Cynicism*, ed. John Corner and Dick Pels (London: Sage, 2003), 96.

25. Ibid.

26. See Erica Seifert, *The Politics of Authenticity in Presidential Campaigns, 1976–2008* (Jefferson, NC: McFarland Press, 2012), 141.

27. Jason Gainous and Kevin M. Wagner, *Tweeting to Power: The Social Media Revolution in American Politics* (New York: Oxford University Press, 2014), 16.

28. Ibid., 13.

29. Michael Barbaro, "Pithy, Mean, and Powerful: How Donald Trump Mastered Twitter for 2016," *New York Times*, October 5, 2015, accessed April 5, 2016, http://www.nytimes.com/2015/10/06/us/politics/donald-trump-twitter-use-campaign-2016.html?_r=0.

30. Antoaneta Roussi, "The Twitter Candidate: Donald Trump's Mastery of Social Media Is His Real Ground Game," *Salon*, February 18, 2016, accessed April 5, 2016, http://www.salon.com/2016/02/18/the_twitter_candidate_donald_trumps_mastery_of_social_media_is_his_real_ground_game/.

31. John Corner and Dick Pels, "Introduction: The Re-Styling of Politics," in *Media and the Restyling of Politics: Consumerism, Celebrity, and Cynicism*, ed. John Corner and Dick Pels (London: Sage, 2003), 8.

32. Liesbet Van Zoonen, *Entertaining the Citizen: When Politics and Popular Culture Converge* (Lanham, MD: Rowman & Littlefield, 2005), 63.

33. Ibid.

34. For an extended treatment of this concept, see George E. Marcus, W. Russell Neuman, and Michael MacKuen, *Affective Intelligence and Political Judgment* (Chicago: University of Chicago Press, 2000).

35. Scammell, *Consumer Democracy*, 157.

36. Ibid., 108.

37. Ibid., 68.

38. Ibid., 137.

39. Henry Jenkins, *Convergence Culture: When Old and New Media Collide* (New York: New York University Press, 2006).

40. Sarah Banet-Weiser, *Authentic TM: The Politics of Ambivalence in a Brand Culture* (New York: New York University Press, 2012).

41. Gabby Kaufman, "Clinton Asks for an Emoji Response to Her Education Plan, with Predictable Results," *Yahoo! Politics*, August 12, 2015, accessed April 5, 2016, https://www.yahoo.com/politics/clinton-asks-for-an-emoji-response-to-her-126539090206.html.

42. Stromer-Galley, *Presidential Campaigning*, 104–139.

43. Cass Sunstein, *Republic.com 2.0* (Princeton, NJ: Princeton University Press, 2009).

44. Gainous and Wagner, *Tweeting to Power*, 152.

45. Nicholas A. John and Shira Dvir-Gvirsman, "'I Don't Like You Any More': Facebook Unfriending by Israelis during the Israel–Gaza Conflict of 2014," *Journal of Communication* 65, no. 6 (2015): 953–974.

46. Magdalena Wojcieszak and Diana C. Mutz, "Online Groups and Political Discourse: Do Online Discussion Spaces Facilitate Exposure to Political Engagement?" *Journal of Communication* 59, no. 1 (2009): 40–56.

47. Gainous and Wagner, *Tweeting to Power*, 124.

48. Eytan Bakshy, Solomon Messing, and Lada Adamic, "Exposure to Ideologically Diverse News and Opinion on Facebook," *Science* 348, no. 6239 (2015): 1130–1132.

49. For example, see Nathan Jurgenson, "Facebook: Fair and Balanced," *Cyborgology*, May 7, 2015, accessed March 25, 2016, https://thesocietypages.org/cyborgology/2015/05/07/facebook-fair-and-balanced/.

50. For example, Stuart Heritage writes in the *Guardian* that "politicized Twitter tends to dissolve into an obnoxious echo chamber, with followers all shouting the same basic viewpoints at each other until everyone gets worked up into such a state of hysteria that they start to believe they're unstoppable." Heritage, "#JezWeCan."

51. Walter Benjamin, "The Work of Art in the Age of Mechanical Reproduction," in *Illuminations*, trans. Harry Zohn (London: Fontana, 1973), 241.

52. Hunter Walker, "Donald Trump Reportedly Paid Actors $50 to Cheer for Him at His 2016 Announcement," *Business Insider*, June 17, 2015, accessed April 5, 2016, http://www.businessinsider.com/paid-actors-at-donald-trump-announcement-2015-6.

53. Lauren Dugan, "Is Newt Gingrich Paying for Thousands of Fake Twitter Followers?" *Adweek*, August 2, 2011, accessed April 5, 2016, http://www.adweek.com/socialtimes/is-newt-gingrich-paying-for-thousands-of-fake-twitter-followers/453661.

54. Marcus, Neuman, and MacKluen, *Affective Intelligence*, 1.

55. Van Zoonen, *Entertaining the Citizen*, 66.

56. John Street, *Politics and Popular Culture* (Philadelphia: Temple University Press), 45–62.

57. "How the 'Fire Big Bird' Meme Could Hurt Mitt Romney," *The Week*, October 5, 2012, accessed August 30, 2014, http://theweek.com/article/index/234411/how-the-fire-big-bird-meme-could-actually-hurtnbspmitt-romney.

58. Devin Dwyer, "Obama TV Ad Uses 'Big Bird' to Mock Romney," *ABC News*, October 9, 2012, accessed August 30, 2014, http://abcnews.go.com/blogs/politics/2012/10/obama-tv-ad-uses-big-bird-to-mock-romney/.

59. Suzi Parker, "Mitt Romney's 'Binders Full of Women,'" *Washington Post*, October 17, 2012, accessed August 30, 2014, http://www.washingtonpost.com/blogs/she-the-people/wp/2012/10/17/mitt-romneys-binders-full-of-women/.

60. Emma Gray, "How 'Nasty Woman' Became A Viral Call for Solidarity," *Huffington Post*, October 20, 2016, accessed January 9, 2017, http://www.huffingtonpost.com/entry/nasty-woman-became-a-call-of-solidarity-for-women-voters_us_5808f6a8e4b02444efa20c92.

61. See Henry Jenkins, "'Why So Socialist?': Unmasking the Joker," *HenryJenkins.org*, August 19, 2009, accessed August 30, 2014, http://henryjenkins.org/2009/08/unmasking_the_joker.html.

62. Eliza Collins, "Poll: Clinton, Trump Most Unfavorable Candidates Ever," *USA Today*, August 31, 2016, accessed January 9, 2017, http://www.usatoday.com/story/news/politics/onpolitics/2016/08/31/poll-clinton-trump-most-unfavorable-candidates-ever/89644296/.

63. Ian Crouch, "Trump: The Man, The Meme," *New Yorker*, December 12, 2015, accessed April 5, 2016, http://www.newyorker.com/culture/culture-desk/trump-man-meme.

64. Jason Koebler, "How r/the_donald Became a Melting Pot of Frustration and Hate," *Vice Motherboard*, July 12, 2016, accessed January 9, 2017, http://motherboard.vice.com/read/what-is-rthedonald-donald-trump-subreddit.

65. Shanto Iyengar and Sean J. Westwood, "Fear and Loathing across Party Lines: New Evidence on Group Polarization," *American Journal of Political Science* 59, no. 3 (2015): 690.

66. Lilliana Mason, "The Rise of Uncivil Agreement: Issue versus Behavioral Polarization in the American Electorate," *American Behavioral Scientist* 57, no. 1 (2013): 155.

67. Ibid.

68. Robert Reich, "The New Tribalism and the Decline of the Nation State," *Huffington Post*, March 24, 2014, accessed August 30, 2014, http://www.huffingtonpost.com/robert-reich/the-new-tribalism-and-the_b_5020469.html.

69. Mason, "The Rise of Uncivil Agreement," 155.

70. For instance, this practice is documented in the Bolivian context in a 2005 documentary film that follows the overseas campaign consultancy work of the U.S. political strategist James Carville. At one point in the film, Carville's team pushes the use of negative ads targeting the opposing candidate, which up until that point had been relatively unfamiliar to Bolivian voters. *Our Brand Is Crisis*, directed by Rachel Boynton (2005; New York: Koch–Lorber Films, 2006), DVD.

71. The Wesleyan Media Project documents the growth of negative campaign advertising in the United States in recent election cycles, showing that the amount of negative ads doubled from roughly 30% of all campaign ads in 2000 to more than 60% of all ads in 2012. See John Wihbey, "Negative Political Ads, the 2012 Campaign and Voter Effects: Research Roundup," *Journalist's Resource*, May 6, 2013, accessed October 5, 2015, http://journalistsresource.org/studies/politics/ads-public-opinion/negative-political-ads-effects-voters-research-roundup#.

72. Sunstein, *Republic.com 2.0*.

73. Kjerstin Thorson, "Facing an Uncertain Reception: Young Citizens and Political Interaction on Facebook," *Information, Communication & Society* 17, no. 2 (2014): 203–216.

74. Kjerstin Thorson, Emily Vraga, and Neta Kligler-Vilenchik, "Don't Push Your Opinions on Me: Young Citizens and Political Etiquette on Facebook," in *Presidential Campaigning and Social Media*, ed. John Allen Hendricks and Dan Schill (New York: Oxford University Press, 2015), 74–93.

75. Nina Eliasoph, *Avoiding Politics: How Americans Produce Apathy in Everyday Life* (Cambridge: Cambridge University Press, 1998), 21.

76. Ibid., 255.

77. Zizi Papacharissi, *A Private Sphere: Democracy in a Digital Age* (Malden, MA: Polity Press, 2010).

78. Chris Perkins, "This Site Makes It Easy to Judge Your Facebook Friends Who Like Trump," *Mashable*, December 9, 2015, accessed April 5, 2016, http://mashable.com/2015/12/09/facebook-friends-like-trump/#helwKYVflgqi.

79. Mark Button and Kevin Mattson, "Deliberative Democracy in Practice: Challenges and Prospects for Civic Deliberation," *Polity* 31, no. 4 (1999): 636.

80. Eliasoph, *Avoiding Politics*, 234.

81. Todd Graham, Daniel Jackson, and Scott Wright, "From Everyday Conversation to Political Action: Talking Austerity in Online 'Third Spaces,'" *European Journal of Communication* 30, no. 6 (2015): 648–665.

82. Peter Dahlgren, *Media and Political Engagement: Citizens, Communication, and Democracy* (Cambridge: Cambridge University Press, 2009).

83. Todd Graham, "Beyond 'Political' Communicative Spaces: Talking Politics on the Wife Swap Discussion Forum," *Journal of Information Technology and Politics* 9, no. 1 (2012): 31–45.

84. Eliasoph, *Avoiding Politics*, 250.

85. Joakim Ekman and Erik Amna, "Political Participation and Civic Engagement: Towards a New Typology," *Human Affairs* 22, no. 3 (2012): 283–300.

86. Ibid., 283.

87. Scammell, *Consumer Democracy*, 157.

88. Schudson, *The Good Citizen*, 294–313.

89. Stromer-Galley, *Presidential Campaigning*, 2–3.

90. For example, see Justin Lewis, Sanna Inthorn, and Karin Wahl-Jorgensen, *Citizens or Consumers?: What the Media Tell Us about Political Participation* (New York, Open University Press, 2005), 138; Nico Carpentier, *Media and Participation: A Site of Ideological-Democratic Struggle* (Bristol, UK: Intellect, 2011), 84.

91. Asa Bennett, "How Labour's Leadership Election Works, and Who Will Win," *The Telegraph,* July 30, 2015, accessed February 12, 2016, http://www.telegraph.co.uk/news/politics/labour/11772723/How-Labours-leadership-election-works-and-who-will-win.html.

92. Nick Anstead and Andrew Chadwick, "Parties, Election Campaigning, and the Internet: Towards a Comparative Institutional Approach," in *The Routledge Handbook of Internet Politics,* ed. Andrew Chadwick and Philip. N. Howard (London: Routledge, 2008), 56–71.

93. See Chantal Mouffe, "Deliberative Democracy or Agonistic Pluralism," *Reihe Politikwissenschaft/ Political Science Series* 72 (2000), accessed February 1, 2016, https://www.ihs.ac.at/publications/pol/pw_72.pdf.

94. Liz Cohen, *A Consumer's Republic: The Politics of Mass Consumption* (New York: Knopf, 2003), 347.

Chapter 5

1. Lee Rainie, Paul Hitlin, Mark Jurkowitz, Michael Dimock, and Shawn Neidorf, "The Viral Kony 2012 Video," *Pew Research Internet Project,* March 15, 2012, accessed November 21, 2014, http://www.pewinternet.org/2012/03/15/the-viral-kony-2012-video/.

2. Jessica Testa, "Two Years after KONY 2012, Has Invisible Children Grown Up?," *Buzzfeed,* March 9, 2014, accessed November 21, 2014, http://www.buzzfeed.com/jtes/two-years-after-kony-2012-has-invisible-children-grown-up.

3. For a summary of the Kony 2012 controversy and Internet parody memes, see "Kony 2012," *Know Your Meme,* accessed November 21, 2014, http://knowyourmeme.com/memes/events/kony-2012.

4. For example, see Christopher Babu, "Viral Fever: Kony and Slacktivism," *Sunday Guardian,* April 8, 2012, accessed November 21, 2014, http://www.sunday-guardian.com/artbeat/viral-fever-kony-a-slacktivism.

5. Quoted in Testa, "Two Years After."

6. For a detailed discussion of the Sandra Bland case and the #SayHerName campaign that followed, see Amanda Sakuma, "The Time Has Come to 'Say Her Name,'" *MSNBC,* August 5, 2015, accessed February 16, 2016, http://www.msnbc.com/msnbc/the-time-has-come-say-her-name.

7. Colin Daileda, "Campaign Highlights Police Brutality against Black Women," *Mashable,* May 26, 2015, accessed February 16, 2016, http://mashable.com/2015/05/26/police-brutality-black-women/#sKEAeuqNlOq5.

8. Tia Oso, "I Am the Black Woman Who Interrupted the Netroots Presidential Town Hall, and This Is Why," *Mic,* July 21, 2015, accessed February 16, 2016, http://mic.com/articles/122629/i-am-the-black-woman-who-interrupted-the-netroots-presidential-town-hall-and-this-is-why#.oXFaY62rN.

9. See Clenora Hudson-Weems, "Resurrecting Emmett Till: The Catalyst of the Modern Civil Rights Movement," *Journal of Black Studies* 29, no. 2 (1998): 179–188.

10. Andrew Chadwick, *The Hybrid Media System: Politics and Power* (New York: Oxford University Press, 2015).

11. Maxwell McCombs and Douglas Shaw, "The Agenda-Setting Function of Mass Media," *Public Opinion Quarterly* 36, no. 2 (1972): 176–187.

12. Sung-Tae Kim and Young-Hwan Lee, "New Functions of Internet Mediated Agenda-Setting: Agenda-Rippling and Reversed Agenda-Setting," *The Korean Journal of Journalism and Communication Studies* 50, no. 3 (2006): 175–204.

13. Yochai Benkler, *The Wealth of Networks: How Social Production Transforms Markets and Freedom* (New Haven, CT: Yale University Press, 2006), 263.

14. Ben Sayre et al., "Agenda Setting in a Digital Age: Tracking Attention to Proposition 8 in Social Media, Online News, and Conventional News," *Policy & Internet* 2, no. 2 (2010): 7–32.

15. Zizi Papacharissi, *Affective Publics: Sentiment, Technology, and Politics* (New York: Oxford University Press, 2014), 48.

16. See Christian Christensen, "Iran: Networked Dissent?" *Counterpunch*, July 2, 2009, accessed November 25, 2014, http://www.counterpunch.org/2009/07/02/iran-networked-dissent/.

17. Papacharissi, *Affective Publics*, 50.

18. Pablo Barbera et al., "The Critical Periphery in the Growth of Social Protests," *PLoS ONE* 10, no. 11 (2015): e0143611, doi:10.1371/journal.pone.0143611.

19. Joel Penney and Caroline Dadas, "(Re)Tweeting in the Service of Protest: Digital Composition and Circulation in the Occupy Wall Street Movement," *New Media & Society* 16, no. 1 (2014): 74–90.

20. David Tewksbury and Jason Rittenberg, *News on the Internet: Information and Citizenship in the 21st Century* (New York: Oxford University Press, 2012), 155.

21. Zizi Papacharissi, *A Private Sphere: Democracy in a Digital Age* (Malden, MA: Polity Press, 2010), 52–53.

22. See Thomas H. Davenport and John C. Beck, *The Attention Economy: Understanding the New Currency of Business* (Cambridge, MA: Harvard Business Press, 2013).

23. Luke Goode, "Social News, Citizen Journalism and Democracy," *New Media & Society* 11, no. 8 (2009): 1295.

24. See Burton Saint John, *Press Professionalization and Propaganda: The Rise of Journalistic Double-Mindedness* (Amherst, NY: Cambria Press, 2010), 77–100.

25. Mark Crispin Miller, introduction to *Propaganda*, by Edward Bernays (New York: IG Publishing, 2005), 15.

26. Papacharissi, *A Private Sphere*, 154.

27. Tewksbury and Rittenberg, *News on the Internet*, 144.

28. Jason Gainous and Kevin M. Wagner, *Tweeting to Power: The Social Media Revolution in American Politics* (New York: Oxford University Press, 2014), 124.

29. Tewksbury and Rittenberg, *News on the Internet*, 178.

30. Peter Dahlgren, "The Internet, Public Spheres, and Political Communication: Dispersion and Deliberation," *Political Communication* 22, no. 2 (2005): 158.

31. Jürgen Habermas, *The Theory of Communicative Action, Volume 1*, trans. Thomas McCarthy (Boston: Beacon Press, 1984).

32. Jürgen Habermas, *Moral Consciousness and Communicative Action*, trans. Shierry Weber Nicholsen and Christian Lenhardt (Cambridge, MA: MIT Press, 1990), 87–89.

33. Edward S. Herman and Noam Chomsky, *Manufacturing Consent: The Political Economy of the Mass Media* (New York: Pantheon Books, 1988).

34. W. Lance Bennett, *News: The Politics of Illusion*, 9th ed. (Chicago: University of Chicago Press, 2011).

35. Simon Sheppard, "American Media, American Bias: The Partisan Press from Broadsheet to Blog" (Ph.D. diss., Johns Hopkins University, 2008).

36. Matthew Levendusky, *How Partisan Media Polarize America* (Chicago: University of Chicago Press, 2013), 4.

37. Bruce A. Williams and Michael X. Delli Carpini, *After Broadcast News: Media Regimes, Democracy, and the New Information Environment* (New York: Cambridge University Press, 2011).

38. Cass Sunstein, *Republic.com 2.0* (Princeton, NJ: Princeton University Press, 2009).

39. "About," *Upworthy*, accessed February 28, 2016, http://www.upworthy.com/about.

40. Eli Pariser, *The Filter Bubble: What the Internet Is Hiding from You* (New York: Penguin Books, 2011).

41. Craig Silverman, "This Analysis Shows how Fake Election News Stories Outperformed Real News on Facebook," *Buzzfeed*, November 16, 2016, accessed January 9, 2017, https://www.buzzfeed.com/craigsilverman/viral-fake-election-news-outperformed-real-news-on-facebook?utm_term=.umNVlmKp6#.jqznYzeNo.

42. See John Street, *Politics and Popular Culture* (Cambridge, MA: Polity, 1997).

43. Bruce Hardy et al., "Stephen Colbert's Civics Lesson: How Colbert Super PAC Taught Viewers about Campaign Finance," *Mass Communication and Society* 17, no. 3 (2014): 329–353.

44. Jonathan Gray, Jeffrey P. Jones, and Ethan Thompson, "The State of Satire, the Satire of State," in *Satire TV: Politics and Comedy in the Post-Network Era*, ed. Jonathan Gray, Jeffrey P. Jones, and Ethan Thompson (New York: New York University Press, 2009), 15.

45. Patrick Kingsley, "Egyptian Satirist Bassem Youssef Winds up TV Show due to Safety Fears," *Guardian*, June 2, 2014, accessed November 22, 2014, http://www.theguardian.com/media/2014/jun/02/bassem-youssef-closes-egyptian-satire-tv-show-over-safety-fears.

46. Brian Montopoli, "Jon Stewart Rally Attracts Estimated 215,000," *CBS News*, October 31, 2010, accessed November 22, 2014, http://www.cbsnews.com/news/jon-stewart-rally-attracts-estimated-215000/.

47. See Goode, "Social News," 1287–1305.

48. C. W. Anderson, "From Indymedia to Demand Media: Journalism's Visions of Its Audience and the Horizons of Democracy," in *The Social Media Reader*, ed. Michael Mandiberg (New York: New York University Press, 2012), 82.

49. Ibid., 88.

50. Fredrick C. Harris, "The Next Civil Rights Movement?" *Dissent*, Summer 2015, accessed March 1, 2016, https://www.dissentmagazine.org/article/black-lives-matter-new-civil-rights-movement-fredrick-harris.

51. Gainous and Wagner, *Tweeting to Power*, 13.

52. Ibid., 105.

53. Marina Fang, "Donald Trump Tweets a Wildly Inaccurate Graphic to Portray Black People as Murderers," *Huffington Post*, November 22, 2015, accessed March 1, 2016, http://www.huffingtonpost.com/entry/donald-trump-inaccurate-tweet_us_56524c0de4b0879a5b0b6c10.

54. Jennifer Stromer-Galley, *Presidential Campaigning in the Internet Age* (New York: Oxford University Press, 2014).

55. Williams and Delli Carpini, *After Broadcast News*.

56. See William Davies, "The Age of Post-Truth Politics," *The New York Times*, August 24, 2016, accessed January 9, 2017, https://www.nytimes.com/2016/08/24/opinion/campaign-stops/the-age-of-post-truth-politics.html?_r=0.

57. Gainous and Wagner, *Tweeting to Power*, 41.

58. See Maria Bakardjieva, "Subactivism: Lifeworld and Politics in the Age of the Internet," *The Information Society* 25 (2009): 91–104; Nico Carpentier, *Media and Participation: A Site of Ideological-Democratic Struggle* (Bristol, UK: Intellect, 2011).

59. Dahlgren, "The Internet, Public Spheres, and Political Communication," 158.

60. Jessica Vitak et al., "It's Complicated: Facebook Users' Political Participation in the 2008 Election," *CyberPsycholology, Behavior and Social Networking* 14, no. 3 (2011): 107–114.

61. Nicole Eliasoph, *Avoiding Politics: How Americans Produce Apathy in Everyday Life* (Cambridge: Cambridge University Press, 1998), 257.

62. Nicholas O'Shaughnessy, "The Death and Life of Propaganda," *Journal of Public Affairs* 12, no. 1 (2012): 29.

63. Hugh Atkin, "Will the Real Mitt Romney Please Stand Up (feat. Eminem)," *YouTube*, March 19, 2012, accessed November 2, 2014, http://www.youtube.com/watch?v=bxch-yi14BE.

64. Anderson, "From Indymedia to Demand Media," 88.

65. Nancy Fraser, "Rethinking the Public Sphere: A Contribution to the Critique of Actually Existing Democracy," *Social Text* 25/26 (1990): 56–80.

66. Manuel Castells, "Communication, Power, and Counter-power in the Network Society," *International Journal of Communication* 1 (2007): 248.

67. Manuel Castells, *Networks of Outrage and Hope: Social Movements in the Internet Age* (Malden, MA: Polity Press, 2012), 194.

68. Gainous and Wagner, *Tweeting to Power*, 13.

69. Chadwick, *The Hybrid Media System*, 4.

70. Tewksbury and Rittenberg, *News on the Internet*, 164.

71. Robert McChesney, *Digital Disconnect: How Capitalism Is Turning the Internet against Democracy* (New York: New Press, 2013), 179.

72. Ibid., 189.

73. "The Evolving Role of News on Twitter and Facebook," Pew Research Center, July 2015, accessed February 1, 2016, http://www.journalism.org/files/2015/07/Twitter-and-News-Survey-Report-FINAL2.pdf.

74. Jonah Berger and Katherine L. Milkman, "What Makes Online Content Viral?," *Journal of Marketing Research* 49, no. 2 (2012): 192–205.

75. Maria Konnikova, "The Six Things That Make Stories Go Viral Will Amaze, and Maybe Infuriate, You," *New Yorker*, January 21, 2014, accessed February 1, 2016, http://www.newyorker.com/tech/elements/the-six-things-that-make-stories-go-viral-will-amaze-and-maybe-infuriate-you.

76. McChesney, *Digital Disconnect*, 187.

77. For example, see Olivia Blair, "Martin Shkreli: 'Most Hated Man on the Internet' Breaks Pledge to Lower Cost of HIV-treating Drug," *Independent (UK)*, November 26, 2015, accessed February 10, 2016, http://www.independent.co.uk/news/people/martin-shkreli-most-hated-man-on-the-internet-breaks-pledge-to-lower-cost-of-hiv-treating-drug-a6750291.html.

78. Matt Egan, "Hillary Clinton Tweet Crushes Biotech Stocks," *CNN*, September 22, 2015, accessed February 10, 2016, http://money.cnn.com/2015/09/21/investing/hillary-clinton-biotech-price-gouging/.

79. For example, see Katie Halper, "These 37 Tweets Show How the Pharma CEO Gouging AIDS Patients Is Even Worse Than You Think," *Raw Story*, September 21, 2015, accessed February 10, 2016, http://www.rawstory.com/2015/09/these-37-tweets-show-how-the-pharma-ceo-gouging-aids-patients-is-even-worse-than-you-think/.

80. For instance, during the period of September 15, 2015, to January 22, 2016, Shkreli was the subject of 46 articles posted by the liberal news site *Raw Story*, 33 posted by *Salon*, 32 by *Gawker*, 30 by *Addicting Info*, and 17 by *Buzzfeed*. By contrast, the U.S. right-wing blogosphere showed relatively little interest in the Shkreli story; looking at a similar sample over the same time frame, *The Daily Caller* posted 6 Shkreli articles, *The Blaze* and *Breitbart* posted 3 each, and *Red State* and *Hot Air* posted only 1 each.

81. For example, see Daniel Villarreal, "This Is Price Gouging Martin Shkreli's Dating Profile," *Unicorn Booty*, September 22, 2015, accessed February 10, 2016, https://unicornbooty.com/this-is-price-gouging-martin-shkrelis-dating-profile/.

82. For example, see Tom Barnes, "Pharma Bro Martin Shkreli Bought the Only Copy of Wu-Tang Clan's New Album for $2 Million," *Mic*, December 9, 2015, accessed February 10, 2016, http://mic.com/articles/130041/pharma-bro-martin-shkreli-bought-the-only-copy-of-wu-tang-clan-s-new-album-for-2-million#.hBAZ8jrj6.

83. Kali Holloway, "Here Are 14 Douchebags We'd Love to See Disappear in 2016," *Raw Story*, December 26, 2015, accessed February 10, 2016, http://www.rawstory.com/2015/12/here-are-14-douchebags-wed-love-to-see-disappear-in-2016/.

84. Tbogg, "Here Are the 11 Most Punchable Faces of 2015," *Raw Story*, December 16, 2015, accessed February 10, 2016, http://www.rawstory.com/2015/12/most-punchable-faces-2015/.

85. Sam Biddle, "Here's Your Martin Shkreli Perp Walk Schadenfreude Gallery," *Gawker*, December 17, 2015, accessed February 10, 2016, http://gawker.com/heres-your-martin-shkreli-perp-walk-schadenfreude-galle-1748514916.

86. Scott Eric Kaufman, "Twitter Nearly Drowns in Schadenfreude at News of 'Pharmo Bro' Martin Shkreli's Arrest Thursday," *Salon*, December 17, 2015, accessed February 10, 2016, http://www.salon.com/2015/12/17/twitter_nearly_drowns_in_schadenfreude_at_news_of_pharmo_bro_martin_shkrelis_arrest_on_thursday/.

87. Castells, *Networks of Outrage and Hope*, 194.

88. Margaret Scammell, *Consumer Democracy: The Marketing of Politics* (Cambridge: Cambridge University Press, 2014).

89. Papacharissi, *Affective Publics*, 26.

90. Ibid., 120.

91. Harold Laswell, "The Theory of Political Propaganda," *The American Political Science Review* 21, no. 3 (1927): 631.

92. Papacharissi, *Affective Publics*, 132–135.

93. Ethan Zuckerman, "Understanding Digital Civics," *My Heart's in Accra*, August 30, 2012, accessed May 1, 2015, http://www.ethanzuckerman.com/blog/2012/08/30/understanding-digital-civics/.

Chapter 6

1. See Charles Tilly, *Popular Contention in Great Britain, 1758–1834* (Cambridge, MA: Harvard University Press, 1995).

2. W. Lance Bennett and Alexandra Segerberg, *The Logic of Connective Action: Digital Media and the Personalization of Contentious Politics* (New York: Cambridge University Press, 2013).

3. Bruce Bimber, Andrew Flanagin, and Cynthia Stohl, *Collective Action in Organizations: Interaction and Engagement in an Era of Technological Change* (Cambridge: Cambridge University Press, 2012), 150–157.

4. Ibid., 187.

5. Peter Dahlgren, *The Political Web: Media, Participation, and Alternative Democracy* (New York: Palgrave, 2013), 63–64.

6. Malcolm Galdwell, "Small Change: Why the Revolution Will Not Be Tweeted," *New Yorker*, October 4, 2010, accessed February 28, 2016, http://www.newyorker.com/magazine/2010/10/04/small-change-3.

7. Henrik Christensen, "Political Activities on the Internet: Slacktivism or Political Participation by Other Means?" *First Monday* 16, no. 2 (2011), accessed February 15, 2016, http://dx.doi.org/10.5210/fm.v16i2.3336.

8. Ibid.

9. Evgeny Morozov, "Foreign Policy: Brave New World of Slacktivism," *NPR*, May 19, 2009, accessed February 28, 2016, http://www.npr.org/templates/story/story.php?storyId=104302141.

10. Natalie Fenton and Veronica Barassi, "Alternative Media and Social Networking Sites: The Politics of Individuation and Political Participation," *The Communication Review* 14, no. 3 (2011): 186–187.

11. See Dahlgren, *The Political Web*, 52.

12. Giovanna Mascheroni, "Performing Citizenship Online: Identity, Subactivism and Participation," *Observatorio* 7, no. 3 (2013): 107.

13. Anastasia Khoo, "How Support for Marriage Equality Went Viral on Facebook," *Huffington Post*, June 22, 2013, accessed October 4, 2014, http://www.huffingtonpost.com/anastasia-khoo/how-support-for-marriage-_b_3483736.html?utm_hp_ref=impact&ir=Impact.

14. My discussion here of "the ladder of engagement" shares certain resonances, as well as divergences, from Sherry Arnstein's widely cited "ladder of citizen participation." Arnstein's

ladder refers to different levels of citizen power in the decision-making activities of institutional bodies, such as government agencies and community boards. For instance, consulting, one of the lower rungs of this ladder included in the category of tokenism, involves citizens being granted the opportunity to provide input, but not with the guarantee that it will have any bearing on subsequent decision-making processes. In a certain sense, the symbolic action of citizen marketing as a whole could be likened to a consulting role, as its practitioners work to both articulate and influence public opinion in ways that may or may not be taken into account by powerful decision makers such as politicians and organizational leaders. At the top of Arnstein's ladder, by contrast, is partnership, delegated power, and citizen control, which indicates a far more active and participatory role for citizens in institutional decision-making processes. To some degree, this highest rung of the ladder could be compared to high-level organizational involvement in political campaigns and advocacy groups, which suggests that moving from symbolic media participation to higher-cost organizational participation could be viewed as moving up Arnstein's ladder. However, this comparison has notable limits. For instance, the kind of high-commitment organizational participation of primary concern in the slacktivism debate does not necessarily involve any enhanced decision-making power on the part of citizens; on the contrary, it can often mean simply serving the needs of organizational leadership in more time-intensive and prolonged ways. Ultimately, the ladder of engagement in political movements and campaigns referred to in the current discussion has less to do with variations in citizen decision-making power than with qualitative differences in how citizens contribute to collective goals—for example, through lower-cost symbolic labor versus higher-cost organizational labor. See Sherry R. Arnstein, "A Ladder of Citizen Participation," *Journal of the American Institute of Planners* 35, no. 4 (July 1969): 216–224.

15. Christensen, "Political Activities on the Internet."

16. Pablo Barbera et al., "The Critical Periphery in the Growth of Social Protests," *PLoS ONE* 10, no. 11 (2015): e0143611, doi:10.1371/journal.pone.0143611.

17. Julie Sloane, "In Social Movements, 'Slacktivists' Matter," *Annenberg School for Communication, University of Pennsylvania*, December 7, 2015, accessed February 15, 2016, https://www.asc.upenn.edu/news-events/news/social-movements-slacktivists-matter.

18. Manuel Castells, "Communication, Power, and Counter-power in the Network Society," *International Journal of Communication* 1 (2007): 249.

19. Michael Schudson, *Advertising, the Uneasy Persuasion: Its Dubious Impact on American Society* (New York: Basic Books, 1984).

20. For example, see Jeffrey P. Jones, *Entertaining Politics: New Political Television and Civic Culture* (Oxford: Rowman & Littlefield, 2005).

21. Barrie Axford and Richard Huggins, "Anti-politics or the Triumph of Postmodern Populism in Promotional Cultures?," *Javnost: The Public 4*, no. 3, (1997): 5–27.

22. Walter Benjamin, "The Work of Art in the Age of Mechanical Reproduction," in *Illuminations*, trans. Harry Zohn (London: Fontana, 1973), 241.

23. Stuart Ewen, *All Consuming Images: The Politics of Style in Contemporary Culture*, revised ed. (New York: Basic Books, 1999).

24. Zizi Papacharissi, *Affective Publics: Sentiment, Technology, and Politics* (New York: Oxford University Press, 2014), 48.

25. C. W. Anderson, "From Indymedia to Demand Media: Journalism's Visions of Its Audience and the Horizons of Democracy," in *The Social Media Reader*, ed. Michael Mandiberg (New York: New York University Press, 2012), 88.

26. Jonah Berger and Katherine L. Milkman, "What Makes Online Content Viral?," *Journal of Marketing Research* 49, no. 2 (2012): 192–205.

27. See John Street, "In Praise of Political Packaging? Political Coverage as Popular Culture," *International Journal of Press/Politics* 1, no. 2 (1996): 126–133. Margaret Scammell also makes a similar argument in her treatment of electoral political marketing, claiming that it has the potential to make politics more accessible and engaging for citizens by tapping

into experiences of emotion and pleasure. Margaret Scammell, *Consumer Democracy: The Marketing of Politics* (Cambridge: Cambridge University Press, 2014).

28. André Jansson, "The Mediatization of Consumption: Towards an Analytical Framework of Image Culture," *Journal of Consumer Culture* 2, no. 1 (2002): 26.

29. Jean Baudrillard, *Simulacra and Simulation* (Ann Arbor, MI: University of Michigan Press, 1994), 6.

30. Jansson, "Mediatization of Consumption," 5.

31. Ibid.

32. Stuart Hall, "Encoding/Decoding," in *Culture, Media, Language*, ed. Stuart Hall et al. (London: Hutchison, 1980), 128–138.

33. Robert M. Bond et al., "A 61-Million-Person Experiment in Social Influence and Political Mobilization," *Nature* 489, no. 7415 (2013): 295–298.

34. Nancy Baym, "Data Not Seen: The Uses and Shortcomings of Social Media Metrics," *First Monday* 18, no. 10 (2013), accessed May 5, 2015, doi:10.5210/fm.v18i10.4873.

35. David Karpf, *The MoveOn Effect: The Unexpected Transformation of American Political Advocacy* (New York: Oxford University Press, 2012), 168.

36. See Bruce Bimber and Richard Davis, *Campaigning Online: The Internet in U.S. Elections* (New York: Oxford University Press, 2003).

37. Lee Rainie, "Digital Divides 2015," *Pew Research Center*, September 22, 2015, accessed February 28, 2016, http://www.pewinternet.org/2015/09/22/digital-divides-2015/.

38. Alexander Van Deursen and Jan Van Dijk, "Internet Skills and the Digital Divide," *New Media & Society* 13, no. 6 (2011): 893–911.

39. Ibid.

40. Robert McChesney, *Digital Disconnect: How Capitalism Is Turning the Internet against Democracy* (New York: New Press, 2013), 216.

41. Howard Rheingold, "Using Participatory Media and Public Voice to Encourage Civic Engagement," in *Civic Life Online: Learning How Digital Media Can Engage Youth*, ed. Lance W. Bennett (Cambridge, MA: MIT Press, 2008), 97.

42. See Jason Gainous and Kevin M. Wagner, *Tweeting to Power: The Social Media Revolution in American Politics* (New York: Oxford University Press, 2014), 124.

43. See Alex Hearn, "Facebook's Changing Standards: From Beheading to Breastfeeding Images," *Guardian*, October 22, 2013, accessed February 28, 2016, http://www.theguardian.com/technology/2013/oct/22/facebook-standards-beheading-breastfeeding-social-networking.

44. For example, see David Bamman, Brendan O'Connor, and Noah Smith, "Censorship and Deletion Practices in Chinese Social Media," *First Monday* 17, no. 3 (2012), accessed February 28, 2016, http://dx.doi.org/10.5210/fm.v17i3.3943.

45. See Philip N. Howard, Seetal D. Agarwal, and Muzammil M. Hussain, "When Do States Disconnect Their Digital Networks? Regime Responses to the Political Uses of Social Media," *The Communication Review* 14, no. 3 (2011): 216–232.

46. For an account of how Micah White and Kalle Lasn started the Occupy movement through *Adbusters*, see Mattathias Schwartz, "Pre-occupied: The Origins and Future of Occupy Wall Street," *New Yorker*, November 28, 2011, accessed February 28, 2016, http://www.newyorker.com/magazine/2011/11/28/pre-occupied.

47. Micah White, "Clicktivism Is Ruining Left Activism," *Guardian*, August 12, 2010, accessed February 28, 2016, http://www.theguardian.com/commentisfree/2010/aug/12/clicktivism-ruining-leftist-activism.

48. See Liz Cohen, *A Consumer's Republic: The Politics of Mass Consumption* (New York: Knopf, 2003).

49. Heather Savigny, *The Problem of Political Marketing* (New York: Continuum, 2008).

50. Bennett and Segerberg, *The Logic of Connective Action*, 180–181.

51. Kalle Lasn, *Culture Jam* (New York: Eagle Brook, 1999), 123.

52. See Christine Harold, "Pranking Rhetoric: 'Culture Jamming' as Media Activism," *Critical Studies in Media Communication* 21, no. 3 (2004): 189–211.

53. Nico Carpentier, *Media and Participation: A Site of Ideological-Democratic Struggle* (Bristol, UK: Intellect, 2011), 131.
54. Ibid., 84–85.
55. See Limor Shifman, *Memes in Digital Culture* (Cambridge, MA: MIT Press, 2013).
56. Nicholas O'Shaughnessy, "The Death and Life of Propaganda," *Journal of Public Affairs* 1, no. 1 (2012): 33–34.

Methodological Appendix

1. Barney Glaser and Anselm. L. Strauss, *The Discovery of Grounded Theory: Strategies for Qualitative Research* (New Brunswick, NJ: Aldine Transaction, 2009).
2. Herbert J. Rubin and Irene S. Rubin, *Qualitative Interviewing: The Art of Hearing Data* (Thousand Oaks, CA: Sage, 1995), 2.
3. Irving Seidman, *Interviewing as Qualitative Research: A Guide for Researchers in Education and the Social Sciences*, 3rd ed. (New York and London: Teachers College Press, 2006), 9.
4. Thomas R. Lindlof and Bryan C. Taylor, *Qualitative Communication Research Methods*, 2nd ed. (Thousand Oaks, CA: Sage, 2002), 178.
5. Rubin and Rubin, *Qualitative Interviewing*, vii.
6. Ibid.
7. See Yvonna S. Lincoln and Egon G. Guba, *Naturalistic Inquiry* (Newbury Park, CA: Sage, 1985).
8. Martyn Hammersley and Paul Atkinson, *Ethnography: Principles in Practice, Third Edition* (London & New York: Routledge, 2007), 107.
9. Glaser and Strauss, *Discovery of Grounded Theory*, 9.
10. Joel Penney, "Visible Identities, Visual Rhetoric: The Self-Labeled Body as a Popular Platform for Political Persuasion," *International Journal of Communication* 6 (2012): 2318–2336.
11. Joel Penney, "Eminently Visible: The Role of T-Shirts in Gay and Lesbian Public Advocacy and Community Building," *Popular Communication* 11, no. 4 (2013): 289–302.
12. Joel Penney, "Motivations for Participating in 'Viral Politics': A Qualitative Case Study of Twitter Users and the 2012 US Presidential Election," *Convergence: The International Journal of Research into New Media Technologies* 22, no. 1 (2016): 71–87.
13. Joel Penney, "Social Media and Symbolic Action: Exploring Participation in the Facebook Red Equal Sign Profile Picture Campaign," *Journal of Computer-Mediated Communication* 20, no. 1 (2015): 52–66.

References

Alcoff, Linda Martín. *Visible Identities: Race, Gender, and the Self.* New York: Oxford University Press, 2005.

Anderson, C. W. "From Indymedia to Demand Media: Journalism's Visions of Its Audience and the Horizons of Democracy." In *The Social Media Reader*, edited by Michael Mandiberg, 77–96. New York: New York University Press, 2012.

Anstead, Nick, and Andrew Chadwick. "Parties, Election Campaigning, and the Internet: Towards a Comparative Institutional Approach." In *The Routledge Handbook of Internet Politics*, edited by Andrew Chadwick and Philip. N. Howard, 56–71. London: Routledge, 2008.

Arendt, Hannah. *The Human Condition.* Chicago: University of Chicago Press, 1958.

Arnstein, Sherry R. "A Ladder of Citizen Participation." *Journal of the American Institute of Planners* 35, no. 4 (1969): 216–224.

Asard, Erik, and W. Lance Bennett, *Democracy and the Marketplace of Ideas: Communication and Government in Sweden and the United States.* Cambridge: Cambridge University Press, 1997.

Axford, Barrie. "The Transformation of Politics or Anti-politics?" In *New Media and Politics*, edited by Barrie Axford and Richard Huggins, 1–29. London: Sage, 2001.

Axford, Barrie, and Richard Huggins. "Anti-politics or the Triumph of Postmodern Populism in Promotional Cultures?" *Javnost: The Public* 4, no. 3 (1997): 5–27.

Bakardjieva, Maria. "Subactivism: Lifeworld and Politics in the Age of the Internet." *The Information Society* 25 (2009): 91–104.

Bakshy, Eytan, Solomon Messing, and Lada Adamic. "Exposure to Ideologically Diverse News and Opinion on Facebook." *Science* 348, no. 6239 (2015): 1130–1132.

Baldwin-Philippi, Jessica. *Using Technology, Building Democracy: Digital Campaigning and the Construction of Citizenship.* New York: Oxford University Press, 2015.

Banet-Weiser, Sarah. *Authentic TM: The Politics of Ambivalence in a Brand Culture.* New York: New York University Press, 2012.

Barbera, Pablo, Ning Wang, Richard Bonneau, John T. Jost, Jonathan Nagler, Joshua Tucker, and Sandra Gonzalez-Bailon. "The Critical Periphery in the Growth of Social Protests." *PLoS ONE* 10, no. 11 (2015), e0143611, doi:10.1371/journal.pone.0143611. Accessed February 1, 2016.

Barnhurst, Kevin. "Politics in the Fine Meshes: Young Citizens, Power, and Media." *Media, Culture, and Society* 20, no. 2 (1998): 201–218.

Baudrillard, Jean. *Simulacra and Simulation.* Ann Arbor, MI: University of Michigan Press, 1994.

Baym, Nancy. "Data Not Seen: The Uses and Shortcomings of Social Media Metrics." *First Monday* 18, no. 10 (2013), doi:10.5210/fm.v18i10.4873. Accessed May 5, 2015.

Benjamin, Walter. "The Work of Art in the Age of Mechanical Reproduction." In *Illuminations*. Translated by Harry Zohn, 217–252. London: Fontana, 1973.

Benkler, Yochai. *The Wealth of Networks: How Social Production Transforms Markets and Freedom*. New Haven, CT: Yale University Press, 2006.

Bennett, W. Lance. "Branded Political Communication: Lifestyle Politics, Logo Campaigns, and the Rise of Global Citizenship." In *The Politics behind Products*, edited by Michele Micheletti, Andreas Follesdal, and Dietlind Stolle, 101–126. New Brunswick, NJ: Transaction Books, 2008.

Bennett, W. Lance. "Changing Citizenship in the Digital Age." In *Civic Life Online: Learning How Digital Media Can Engage Youth*, edited by Lance W. Bennett, 1–24. Cambridge, MA: MIT Press, 2008.

Bennett, W. Lance. "New Media Power." In *Contesting Media Power*, edited by Nick Couldry and John Curran, 17–37. Lanham, MD: Rowman & Littlefield, 2003.

Bennett, W. Lance. *News: The Politics of Illusion*, 9th ed. Chicago: University of Chicago Press, 2011.

Bennett, W. Lance, and Alexandra Segerberg. *The Logic of Connective Action: Digital Media and the Personalization of Contentious Politics*. New York: Cambridge University Press, 2013.

Berger, Jonah, and Katherine L. Milkman. "What Makes Online Content Viral?" *Journal of Marketing Research* 49, no. 2 (2012): 192–205.

Berlant, Lauren, and Elizabeth Freeman. "Queer Nationality." In *Fear of a Queer Planet*, edited by Michael Warner, 193–229. Minneapolis: University of Minnesota Press, 1993.

Best, Steven, and Douglas Kellner. *The Postmodern Turn*. New York: Guilford Press, 1997.

Beyer, Jessica. *Expect Us: Online Communities and Political Mobilization*. New York: Oxford University Press, 2014.

Bimber, Bruce, and Richard Davis. *Campaigning Online: The Internet in U.S. Elections*. New York: Oxford University Press, 2003.

Bimber, Bruce, Andrew Flanagin, and Cynthia Stohl. *Collective Action in Organizations: Interaction and Engagement in an Era of Technological Change*. Cambridge: Cambridge University Press, 2012.

Bond, Robert M., Christopher J. Fariss, Jason J. Jones, Adman D. I. Kramer, Cameron Marlow, Jamie E. Settle, and James H. Fowler. "A 61-Million-Person Experiment in Social Influence and Political Mobilization." *Nature* 489, no. 7415 (2012): 295–298.

Boulianne, Shelley. "Social Media Use and Participation: A Meta-Analysis of Current Research." *Information, Communication, and Society* 18, no. 5 (2015): 524–538.

boyd, danah. "Why Youth (Heart) Social Network Sites: The Role of Networked Publics in Teenage Social Life." In *Youth, Identity, and Digital Media*, edited by David Buckingham, 119–142. Cambridge, MA: MIT Press, 2008.

Brighenti, Andrea. *Visibility in Social Theory and Social Research*. New York: Palgrave Macmillan, 2010.

Brown, Ruth E. "Citizen Marketing." In *Handbook of Research on Social Interaction Technologies and Collaboration Software: Concepts and Trends: Concepts and Trends*, edited by Tatyana Dumova, 45–55. Hershey, PA: IGI Global, 2010.

Button, Mark, and Kevin Mattson. "Deliberative Democracy in Practice: Challenges and Prospects for Civic Deliberation." *Polity* 31, no. 4 (1999): 609–637.

Canclini, Nestor Garcia. *Consumers and Citizens: Globalization and Multicultural Conflicts*. Minneapolis: University of Minnesota Press, 2001.

Castells, Manuel. "Communication, Power, and Counter-power in the Network Society." *International Journal of Communication* 1 (2007): 238–266.

Castells, Manuel. *Networks of Outrage and Hope: Social Movements in the Internet Age*. Malden, MA: Polity Press, 2012.

Chadwick, Andrew. *The Hybrid Media System: Politics and Power*. New York: Oxford University Press, 2015.

Chaffee, Lyman. *Political Protest and Street Art: Popular Tools for Democratization in Hispanic Countries*. Westport, CT: Greenwood Press, 1993.

Chanter, Tina. "A Critique of Martín Alcoff's Identity Politics: On Power and Universality." *Philosophy Today* 53 (2009): 44–58.

Chasin, Alexandra. *Selling Out: The Gay and Lesbian Movement Goes to Market.* New York: Palgrave, 2000.

Christensen, Henrik. "Political Activities on the Internet: Slacktivism or Political Participation by Other Means?" *First Monday* 16, no. 2 (2011), http://dx.doi.org/10.5210/fm.v16i2.3336. Accessed February 15, 2015.

Cialdini, Robert. "Harnessing the Science of Persuasion." *Harvard Business Review* 79 (2001): 72–79.

Cohen, Liz. *A Consumer's Republic: The Politics of Mass Consumption.* New York: Knopf, 2003.

Corner, John, and Dick Pels. "Introduction: The Re-Styling of Politics." In *Media and the Restyling of Politics: Consumerism, Celebrity, and Cynicism,* edited by John Corner and Dick Pels, 1–18. London: Sage, 2003.

Crane, Diane. *Fashion and Its Social Agendas: Class, Gender, and Identity in Clothing.* Chicago: University of Chicago Press, 2000.

Crenshaw, Kimberlé Williams. "Mapping the Margins: Intersectionality, Identity Politics, and Violence against Women of Color." *Stanford Law Review* 43, no. 6 (1991): 1241–1299.

Dahlberg, Lincoln. "Re-constructing Digital Democracy: An Outline of Four 'Positions.'" *New Media & Society* 13, no. 6 (2011): 855–872.

Dahlgren, Peter. "The Internet, Public Spheres, and Political Communication: Dispersion and Deliberation." *Political Communication* 22, no. 2 (2005): 147–162.

Dahlgren, Peter. "Media, Citizenship, and Civic Culture." In *Mass Media and Society,* edited by James Curran and Michael Gurevitch, 310–328. London and New York: Arnold, 2000.

Dahlgren, Peter. *Media and Political Engagement: Citizens, Communication, and Democracy.* Cambridge: Cambridge University Press, 2009.

Dahlgren, Peter. *The Political Web: Media, Participation, and Alternative Democracy.* New York: Palgrave, 2013.

Danesi, Marcel. *Why It Sells: Decoding the Meanings of Brand Names, Logos, Ads, and Other Marketing and Advertising Ploys.* New York: Rowman & Littlefield, 2007.

Davenport, Thomas H., and John C. Beck. *The Attention Economy: Understanding the New Currency of Business.* Cambridge, MA: Harvard Business Press, 2013.

Davis, Susan. *Parades and Power: Street Theatre in Nineteenth-Century Philadelphia.* Berkeley: University of California Press, 1988.

Davison, Patrick. "The Language of Internet Memes." In *The Social Media Reader,* edited by Michael Mandiberg, 120–135. New York: New York University Press, 2012.

Dean, Jodi. "Claiming a Division, Naming a Wrong." *Theory & Event* 14, no. 4 (2011): Supplement.

DeLuca, Kevin. "Unruly Arguments: The Body Rhetoric of Earth First!, ACT UP, and Queer Nation." *Argumentation and Advocacy* 36, no. 1 (1999): 9–21.

DeLuca, Kevin, and Jennifer Peeples. "From Public Sphere to Public Screen: Democracy, Activism, and the 'Violence' of Seattle." *Critical Studies in Media Communication* 19, no. 2 (2002): 12–151.

Douglas, Mary. *Purity and Danger.* New York: Praeger, 1966.

Durkheim, Emile. *The Elementary Forms of Religious Life.* Translated by Joseph Ward Swain. New York: Free Press, 1965.

Earl, Jennifer, and Katrina Kimport. *Digitally Enabled Social Change: Activism in the Internet Age.* Cambridge, MA: MIT Press, 2011.

Edwards, Richard L. and Chuck Tryon. "Political Video Mashups as Allegories of Citizen Empowerment." *First Monday* 14, no. 10 (2009), http://dx.doi.org/10.5210/fm.v14i10.2617. Accessed February 2, 2016.

Ekman, Joakim, and Erik Amna. "Political Participation and Civic Engagement: Towards a New Typology." *Human Affairs* 22, no. 3 (2012): 283–300.

Eliasoph, Nina. *Avoiding Politics: How Americans Produce Apathy in Everyday Life.* Cambridge: Cambridge University Press, 1998.

Endersby, James, and Michael Towle. "Tailgate Partisanship: Political and Social Expression through Bumper Stickers." *Social Science Journal* 33, no. 3 (1996): 307–319.

Ewen, Stuart. *All Consuming Images: The Politics of Style in Contemporary Culture*, revised ed. New York: Basic Books, 1999.

Fenton, Natalie, and Veronica Barassi. "Alternative Media and Social Networking Sites: The Politics of Individuation and Political Participation." *The Communication Review* 14, no. 3 (2011): 179–196.

Fischer, Roger A. *Tippecanoe and Trinkets Too: The Material Culture of American Presidential Campaigns 1828–1984*. Urbana and Chicago: University of Illinois Press, 2008.

Fraser, Nancy. "Rethinking the Public Sphere: A Contribution to the Critique of Actually Existing Democracy." *Social Text* 25/26 (1990): 56–80.

Gainous, Jason, and Kevin M. Wagner. *Tweeting to Power: The Social Media Revolution in American Politics*. New York: Oxford University Press, 2014.

Gibbins, John, and Bo Reimer. *The Politics of Postmodernity: An Introduction to Contemporary Politics and Culture*. London: Sage, 1999.

Gibson, Rachel K. "Party Change, Social Media and the Rise of 'Citizen-Initiated Campaigning.'" *Party Politics* 21, no. 2 (2015): 183–197.

Giddens, Anthony. *Modernity and Self-Identity: Self and Society in the Late Modern Age*. Stanford, CA: Stanford University Press, 1991.

Glaser, Barney, and Anselm. L. Strauss. *The Discovery of Grounded Theory: Strategies for Qualitative Research*. New Brunswick, NJ: Aldine Transaction, 2009.

Gluckman, Amy, and Betsy Reed. *Homo Economics: Capitalism, Community, and Lesbian and Gay Life*. New York: Routledge, 1997.

Goode, Luke. "Social News, Citizen Journalism and Democracy." *New Media & Society* 11, no. 8 (2009): 1287–1305.

Graham, Todd. "Beyond 'Political' Communicative Spaces: Talking Politics on the Wife Swap Discussion Forum." *Journal of Information Technology and Politics* 9, no. 1 (2012): 31–45.

Graham, Todd, Daniel Jackson, and Scott Wright. "From Everyday Conversation to Political Action: Talking Austerity in Online 'Third Spaces.'" *European Journal of Communication* 30, no. 6 (2015): 648–665.

Gray, Jonathan, Jeffrey P. Jones, and Ethan Thompson. "The State of Satire, the Satire of State." In *Satire TV: Politics and Comedy in the Post-Network Era*, edited by Jonathan Gray, Jeffrey P. Jones, and Ethan Thompson, 3–36. New York: New York University Press, 2009.

Gross, Larry. *Up from Invisibility: Lesbians, Gay Men, and the Media in America*. New York: Columbia University Press, 2001.

Gustafsson, Nils. "This Time It's Personal: Social Networks, Viral Politics and Identity Management." In *Emerging Practices in Cyberculture and Social Networking*, edited by Daniel Riha and Anna Maj, 3–23. New York: Rodopi, 2010.

Habermas, Jürgen. *Moral Consciousness and Communicative Action*. Translated by Shierry Weber Nicholsen and Christian Lenhardt. Cambridge, MA: MIT Press, 1990.

Habermas, Jürgen. *Structural Transformation of the Public Sphere*. Translated by Thomas Burger. Cambridge, MA: MIT Press, 1989.

Habermas, Jürgen. *The Theory of Communicative Action, Volume 1*. Translated by Thomas McCarthy. Boston: Beacon Press, 1984.

Hall, Stuart. "What Is This 'Black' in Black Popular Culture?" In *Stuart Hall: Critical Dialogues in Cultural Studies*, edited by David Morley and Kuan-Hsing Chen, 468–478. London: Routledge, 1996.

Hall, Stuart. "Encoding/Decoding." In *Culture, Media, Language*, edited by Stuart Hall, Dorothy Hobson, Andrew Love, and Paul Willis, 128–138. London: Hutchison, 1980.

Hammersley, Martyn, and Paul Atkinson. *Ethnography: Principles in Practice, Third Edition*. London and New York: Routledge, 2007.

Hardy, Bruce, Jeffrey Gottfried, Kenneth Winneg, and Kathleen Hall Jamieson. "Stephen Colbert's Civics Lesson: How Colbert Super PAC Taught Viewers about Campaign Finance." *Mass Communication and Society* 17, no. 3 (2014): 329–353.

Harold, Christine. "Pranking Rhetoric: 'Culture Jamming' as Media Activism." *Critical Studies in Media Communication* 21, no. 3 (2004): 189–211.

Harris, Bob. *Politics and the Rise of the Press: Britain and France, 1620–1800.* London: Routledge, 1996.

Hearn, Alison. "Brand Me 'Activist.'" In *Commodity Activism: Cultural Resistance in Neoliberal Times*, edited by Roopali Mukherjee and Sarah Banet-Weiser, 23–38. New York: New York University Press, 2012.

Hebdige, Dick. *Subculture: The Meaning of Style.* London: Methuen, 1979.

Herman, Edward S., and Noam Chomsky. *Manufacturing Consent: The Political Economy of the Mass Media.* New York: Pantheon Books, 1988.

Hosch-Dayican, Bengu, Chintan Amrit, Kees Aarts, and Adrie Dassen. "How Do Online Citizens Persuade Fellow Voters? Using Twitter during the 2012 Dutch Parliamentary Election Campaign." *Social Science Computer Review* 34, no. 2 (2016): 135–152.

Howard, Philip N., Seetal D. Agarwal, and Muzammil M. Hussain. "When Do States Disconnect Their Digital Networks? Regime Responses to the Political Uses of Social Media." *The Communication Review* 14, no. 3 (2011): 216–232.

Huba, Jackie, and Ben McConnell. *Citizen Marketers: When People Are the Message.* Chicago: Kaplan Business, 2007.

Huggins, Richard. "The Transformation of the Political Audience?" In *New Media and Politics*, edited by Barrie Axford and Richard Huggins, 127–150. London: Sage, 2001.

Hunter, Christopher. "Political Privacy and Online Politics: How E-Campaigning Threatens Voter Privacy." *First Monday* 7, no. 2 (2002), http://firstmonday.org/issues/issue7_2/hunter/index.html. Accessed August 5, 2015.

Inglis, Ian. "The Continuing Story of John Lennon." *Critical Studies in Media Communication* 22, no. 5 (2005): 451–455.

Iton, Richard. *In Search of the Black Fantastic: Politics and Popular Culture in the Post–Civil Rights Era.* Oxford: Oxford University Press, 2008.

Iyengar, Shanto and Sean J. Westwood. "Fear and Loathing across Party Lines: New Evidence on Group Polarization." *American Journal of Political Science* 59, no. 3 (2015): 690–707.

Jamieson, Kathleen Hall. *Packaging the Presidency: A History and Criticism of Presidential Campaign Advertising.* New York: Oxford University Press, 1984.

Jansson, André. "The Mediatization of Consumption: Towards an Analytical Framework of Image Culture." *Journal of Consumer Culture* 2, no. 1 (2002): 5–31.

Jenkins, Henry. "If It Doesn't Spread, It's Dead (Part One): Media Viruses and Memes." Henryjenkins.org, February 11, 2009, http://henryjenkins.org/2009/02/if_it_doesnt_spread_its_dead_p.html. Accessed March 15, 2015.

Jenkins, Henry. *Convergence Culture: When Old and New Media Collide.* New York: New York University Press, 2006.

Jenkins, Henry, Sam Ford, and Joshua Green. *Spreadable Media: Creating Value and Meaning in a Networked Culture.* New York: New York University Books, 2013.

Jenkins, Henry, Mizuko Ito, and danah boyd. *Participatory Culture in a Networked Era: A Conversation on Youth, Learning, Commerce, and Politics.* New York: Polity Press, 2015.

Jhally, Sut. *The Codes of Advertising: Fetishism and the Political Economy of Meaning in the Consumer Society.* New York: Routledge, 1990.

John, Nicholas A., and Shira Dvir-Gvirsman. "'I Don't Like You Any More': Facebook Unfriending by Israelis during the Israel–Gaza Conflict of 2014." *Journal of Communication* 65, no. 6 (2015): 953–974.

Jones, Jeffery P. *Entertaining Politics: New Political Television and Civic Culture.* Oxford: Rowman & Littlefield, 2005.

Karpf, David. *The MoveOn Effect: The Unexpected Transformation of American Political Advocacy.* New York: Oxford University Press, 2012.

Karpf, David. "Online Political Mobilization from the Advocacy Group's Perspective: Looking Beyond Clicktivism." *Policy & Internet* 2, no. 4 (2010): 7–41.

Katz, Elihu, and Paul Lazarsfeld. *Personal Influence: The Part Played by People in the Flow of Mass Communications*. Glencoe, IL: Free Press, 1965.

Khamis, Sahar, and Katherine Vaughn. "Cyberactivism in the Egyptian Revolution: How Civic Engagement and Citizen Journalism Tilted the Balance." *Arab Media and Society* 13, no. 3 (2011): 1–25.

Kim, Sung-Tae, and Young-Hwan Lee. "New Functions of Internet Mediated Agenda-setting: Agenda-Rippling and Reversed Agenda-Setting." *The Korean Journal of Journalism and Communication Studies* 50, no. 3 (2006): 175–204.

Klein, Naomi. *No Logo*. New York, Picador, 2000.

Kreiss, Daniel. *Taking Our Country Back: The Crafting of Networked Politics from Howard Dean to Barack Obama*. New York: Oxford University Press, 2012.

Kreiss, Daniel, and Philip N. Howard. "New Challenges to Political Privacy: Lessons from the First US Presidential Race in the Web 2.0 Era." *International Journal of Communication* 4 (2010): 1032–1050.

Kristofferson, Kirk, Katherine White, and John Peloza. "The Nature of Slacktivism: How the Social Observability of an Initial Act of Token Support Affects Subsequent Prosocial Action." *Journal of Consumer Research* 40, no. 6 (2014): 1149–1166.

Laclau, Ernesto. *Politics and Ideology in Marxist Theory*. London: New Left Books, 1977.

Lasn, Kalle. *Culture Jam*. New York, Eagle Brook, 1999.

Laswell, Harold. "The Theory of Political Propaganda." *The American Political Science Review* 21, no. 3 (1927): 627–631.

Lees-Marshment, Jennifer. "The Marriage of Politics and Marketing." *Political Studies* 49, no. 4 (2001): 692–713.

Levendusky, Matthew. *How Partisan Media Polarize America*. Chicago: University of Chicago Press, 2013.

Lewis, Justin, Sanna Inthorn, and Karin Wahl-Jorgensen. *Citizens or Consumers?: What the Media Tell Us about Political Participation*. New York: Open University Press, 2005.

Lincoln, Yvonna S., and Egon G. Guba. *Naturalistic Inquiry*. Newbury Park, CA: Sage, 1985.

Lindlof, Thomas R., and Bryan C. Taylor. *Qualitative Communication Research Methods, Second Edition*. Thousand Oaks, CA: Sage, 2002.

Loader, Brian, and Dan Mercea. "Networking Democracy? Social Media Innovations and Participatory Politics." *Information, Communication, & Society* 14, no. 6 (2011): 757–769.

Mangold, W. Glynn, and David Faulds. "Social Media: The New Hybrid Element of the Promotion Mix." *Business Horizons* 52, no. 4 (2009): 357–365.

Marcus, George E., W. Russell Neuman, and Michael MacKuen. *Affective Intelligence and Political Judgment*. Chicago: University of Chicago Press, 2000.

Martin, Richard, and Harold Koda. "V-O-T-E: The 1992 Election in T-shirts." *Textile and Text* 14, no. 3 (1992): 30–35.

Mascheroni, Giovanna. "Performing Citizenship Online: Identity, Subactivism and Participation." *Observatorio* 7, no. 3 (2013): 93–119.

Mason, Lilliana. "The Rise of Uncivil Agreement: Issue versus Behavioral Polarization in the American Electorate." *American Behavioral Scientist* 57, no. 1 (2013): 140–159.

McChesney, Robert. *Digital Disconnect: How Capitalism Is Turning the Internet against Democracy*. New York: New Press, 2013.

McCombs, Maxwell, and Douglas Shaw. "The Agenda-Setting Function of Mass Media." *Public Opinion Quarterly* 36, no. 2 (1972): 176–187.

McGinnis, Joe. *The Selling of the President 1968*. New York: Trident Press, 1969.

Miller, Mark Crispin. Introduction to *Propaganda*, by Edward Bernays, 9–33. New York: IG Publishing, 2005.

Mor, Yifat, Neta Kligler-Vilenchik, and Ifat Maoz. "Political Expression on Facebook in a Context of Conflict: Dilemmas and Coping Strategies of Jewish-Israeli Youth." *Social Media + Society* 1, no. 2 (2015), doi:10.1177/2056305115606750. Accessed March 15, 2016.

Morozov, Evgeny. "Foreign Policy: Brave New World of Slacktivism." *NPR*, May 19, 2009, http://www.npr.org/templates/story/story.php?storyId=104302141. Accessed February 28, 2016.

Morris, Charles, and John Sloop. "'What Lips These Lips Have Kissed': Refiguring the Politics of Queer Public Kissing." *Communication and Critical/Cultural Studies* 3, no. 1 (2006): 1–26.

Mouffe, Chantal. "Deliberative Democracy or Agonistic Pluralism." *Reihe Politikwissenschaft/ Political Science Series* 72 (2000), https://www.ihs.ac.at/publications/pol/pw_72.pdf. Accessed February 1, 2016.

O'Shaughnessy, Nicholas. "The Death and Life of Propaganda." *Journal of Public Affairs* 12, no. 1 (2012): 29–36.

O'Shaughnessy, Nicholas. *The Phenomenon of Political Marketing*. New York: St. Martin's Press, 1990.

Packard, Vance. *The Hidden Persuaders*. New York: McKay, 1957.

Papacharissi, Zizi. *Affective Publics: Sentiment, Technology, and Politics*. New York: Oxford University Press, 2014.

Papacharissi, Zizi. *A Private Sphere: Democracy in a Digital Age*. Malden, MA: Polity Press, 2010.

Pariser, Eli. *The Filter Bubble: What the Internet Is Hiding from You*. New York: Penguin Books, 2011.

Parkins, Wendy. "'The Epidemic of Purple, White, and Green': Fashion and the Suffragette Movement in Britain 1908–1914." In *Fashioning the Body Politic: Gender, Dress, Citizenship*, edited by Wendy Parkins, 97–124. Oxford: Berg, 2002.

Pasek, Josh, Eian More, and Daniel Romer. "Realizing the Social Internet? Online Social Networking Meets Offline Civic Engagement." *Journal of Information Technology & Politics* 6, no. 3–4 (2009): 197–215.

Penney, Joel. "Motivations for Participating in 'Viral Politics': A Qualitative Case Study of Twitter Users and the 2012 US Presidential Election." *Convergence: The International Journal of Research into New Media Technologies* 22, no. 1 (2016): 71–87.

Penney, Joel. "Social Media and Symbolic Action: Exploring Participation in the Facebook Red Equal Sign Profile Picture Campaign." *Journal of Computer-Mediated Communication* 20, no. 1 (2015): 52–66.

Penney, Joel. "Eminently Visible: The Role of T-Shirts in Gay and Lesbian Public Advocacy and Community Building." *Popular Communication* 11, no. 4 (2013): 289–302.

Penney, Joel. "Visible Identities, Visual Rhetoric: The Self-Labeled Body as a Popular Platform for Political Persuasion." *International Journal of Communication* 6 (2012): 2318–2336.

Penney, Joel, and Caroline Dadas. "(Re)Tweeting in the Service of Protest: Digital Composition and Circulation in the Occupy Wall Street Movement." *New Media & Society* 16, no. 1 (2014): 74–90.

Rheingold, Howard. "Using Participatory Media and Public Voice to Encourage Civic Engagement." In *Civic Life Online: Learning How Digital Media Can Engage Youth*, edited by Lance W. Bennett, 97–118. Cambridge, MA: MIT Press, 2008.

Rheingold, Howard. *Smart Mobs: The Next Social Revolution*. Cambridge, MA: Basic Books, 2002.

Rubin, Herbert J., and Irene S. Rubin. *Qualitative Interviewing: The Art of Hearing Data*. Thousand Oaks, CA: Sage, 1995.

Rushkoff, Douglas. *Media Virus! Hidden Agendas in Popular Culture*. New York: Ballantine Books, 1994.

Saint John, Burton. *Press Professionalization and Propaganda: The Rise of Journalistic Double-Mindedness*. Amherst, NY: Cambria Press, 2010.

Savigny, Heather. *The Problem of Political Marketing*. New York: Continuum, 2008.

Sawer, Marian. "Wearing Your Politics on Your Sleeve: The Role of Political Colours in Social Movements." *Social Movement Studies: Journal of Social, Cultural, and Political Protest* 6, no. 1 (2007): 39–56.

Sayre, Ben, Letitia Bode, Dhavan Shah, Dave Wilcox, and Chirag Shah. "Agenda Setting in a Digital Age: Tracking Attention to Proposition 8 in Social Media, Online News, and Conventional News." *Policy & Internet* 2, no. 2 (2010): 7–32.

Scammell, Margaret. *Consumer Democracy: The Marketing of Politics*. Cambridge: Cambridge University Press, 2014.

Schudson, Michael. *The Good Citizen: A History of American Civic Life*. Cambridge, MA: Harvard University Press, 1998.

Schudson, Michael. *Advertising, the Uneasy Persuasion: Its Dubious Impact on American Society*. New York: Basic Books, 1984.

Shulman, Stuart. "The Case against Mass E-mails: Perverse Incentives and Low Quality Public Participation in U.S. Federal Rulemaking." *Policy & Internet* 1, no. 1 (2009): 23–53.

Seidman, Irving. *Interviewing as Qualitative Research: A Guide for Researchers in Education and the Social Sciences*, 3rd ed. New York and London: Teachers College Press, 2006.

Seifert, Erica. *The Politics of Authenticity in Presidential Campaigns, 1976–2008*. Jefferson, NC: McFarland Press, 2012.

Sender, Katherine. *Business, Not Politics: The Making of the Gay Market*. New York: Columbia University Press, 2004.

Serazio, Michael. "Managing the Digital News Cyclone: Power, Participation, and Political Production Strategies." *International Journal of Communication* 9 (2015): 1907–1925.

Sheppard, Simon. "American Media, American Bias: The Partisan Press from Broadsheet to Blog." Ph.D. diss., Johns Hopkins University, 2008.

Sheridan, David, Jim Ridolfo, and Anthony Michel. *The Available Means of Persuasion*. Anderson, SC: Parlor, 2012.

Shifman, Limor. *Memes in Digital Culture*. Cambridge, MA: MIT Press, 2013.

Shifman, Limor. "Memes in a Digital World: Reconciling with a Conceptual Troublemaker." *Journal of Computer-Mediated Communication* 18, no. 3 (2013): 362–377.

Street, John. "The Celebrity Politician: Political Style and Popular Culture." In *Media and the Restyling of Politics: Consumerism, Celebrity, and Cynicism*, edited by John Corner and Dick Pels, 85–98. London: Sage, 2003.

Street, John. *Politics and Popular Culture*. Philadelphia: Temple University Press, 1997.

Street, John. "In Praise of Political Packaging? Political Coverage as Popular Culture." *International Journal of Press/Politics* 1, no. 2 (1996): 126–133.

Stromer-Galley, Jennifer. *Presidential Campaigning in the Internet Age*. New York: Oxford University Press, 2014.

Sunstein, Cass. *Republic.com 2.0*. Princeton, NJ: Princeton University Press, 2009.

Tannenbaum, Melanie. "Will Changing Your Profile Picture Do Anything for Marriage Equality?" *Scientific American*, March 28, 2013, http://blogs.scientificamerican.com/psysociety/2013/03/28/marriage-equality-and-social-proof/. Accessed April 4, 2015.

Taylor, Philip M. *Munitions of the Mind: A History of Propaganda from the Ancient World to the Present Day*, 3rd ed. Manchester, UK: Manchester University Press, 2003.

Tewksbury, David, and Jason Rittenberg. *News on the Internet: Information and Citizenship in the 21st Century*. New York: Oxford University Press, 2012.

Thompson, Elaine. "Political Culture." In *Americanisation and Australia*, edited by Paul Bell and Roger Bell, 107–122. Sydney: University of New South Wales Press, 1998.

Thompson, John. *The Media and Modernity: A Social Theory of the Media*. Cambridge, MA: Polity Press, 1995.

Thorson, Kjerstin. "Facing an Uncertain Reception: Young Citizens and Political Interaction on Facebook." *Information, Communication & Society* 17, no. 2 (2014): 203–216.

Thorson, Kjerstin, Emily Vraga, and Neta Kligler-Vilenchik. "Don't Push Your Opinions on Me: Young Citizens and Political Etiquette on Facebook." In *Presidential Campaigning and Social Media*, edited by John Allen Hendricks and Dan Schill, 74–93. New York: Oxford University Press, 2015.

Tilly, Charles. *Popular Contention in Great Britain, 1758–1834*. Cambridge, MA: Harvard University Press, 1995.

Van Deursen, Alexander, and Jan Van Dijk. "Internet Skills and the Digital Divide." *New Media & Society* 13, no. 6 (2011): 893–911.

Van Zoonen, Liesbet. *Entertaining the Citizen: When Politics and Popular Culture Converge*. Lanham, MD: Rowman & Littlefield, 2005.

Vitak, Jessica, Apul Zube, Andrew Smock, Caleb T. Carr, Nicole Ellison, and Cliff Lampe. "It's Complicated: Facebook Users' Political Participation in the 2008 Election." *CyberPsycholology, Behavior and Social Networking* 14, no. 3 (2011): 107–114.

Walters, Suzanna. *All the Rage: The Story of Gay Visibility in America*. Chicago: University of Chicago Press, 2001.

Warner, Jamie. "Political Culture Jamming: The Dissident Humor of 'The Daily Show with Jon Stewart.'" *Popular Communication* 5, no. 1 (2007): 17–36.

Warner, Michael. "Publics and Counterpublics." *Public Culture* 14, no. 1 (2002): 49–90.

Warner, Sarah. *Acts of Gaiety: LGBT Performance and the Politics of Pleasure*. Ann Arbor: University of Michigan Press, 2012.

Wernick, Andrew. *Promotional Culture: Advertising, Ideology, and Symbolic Expression*. London: Sage, 1991.

Williams, Bruce A., and Michael X. Delli Carpini. *After Broadcast News: Media Regimes, Democracy, and the New Information Environment*. New York: Cambridge University Press, 2011.

Wilson, Elizabeth. *Adorned in Dreams: Fashion and Modernity*. New Brunswick, NJ: Rutgers University Press.

Wojcieszak, Magdalena, and Diana C. Mutz. "Online Groups and Political Discourse: Do Online Discussion Spaces Facilitate Exposure to Political Engagement?" *Journal of Communication* 59, no. 1 (2009): 40–56.

Wright, Scott. "Populism and Downing Street E-Petitions: Connective Action, Hybridity, and the Changing Nature of Organizing." *Political Communication* 32, no. 3, (2015): 414–433.

Wright, Scott. "'Success' and Online Political Participation: The Case of Downing Street E-petitions." *Information, Communication, & Society* 19, no. 6 (2016): 843–857.

Young, Sally. *The Persuaders: Inside the Hidden Machine of Political Advertising*. Melbourne: Pluto Press Australia, 2004.

Zuckerman, Ethan. "Understanding Digital Civics." *My Heart's in Accra*, August 30, 2012, http://www.ethanzuckerman.com/blog/2012/08/30/understanding-digital-civics/. Accessed May 1, 2015.

Index